SUSTAINING RUSSIA'S ARCTIC CITIES

Studies in the Circumpolar North

Editors:
Olga Ulturgasheva, *University of Manchester*
Alexander D. King, *University of Aberdeen*

The Circumpolar North encapsulates all the major issues confronting the world today: enduring colonial legacies for indigenous people and the landscape, climate change and resource extraction industries, international diplomatic tensions, and lived realities of small communities in the interconnected modern world system. This book series provides a showcase for cutting-edge academic research on the lives of Arctic and Sub-arctic communities past and present. Understanding the contemporary Circumpolar North requires a multiplicity of perspectives and we welcome works from the social sciences, humanities and the arts.

Volume 1
Leaving Footprints in the Taiga
Luck, Spirits and Ambivalence among the Siberian Orochen Reindeer Herders and Hunters
Donatas Brandišauskas

Volume 2
Sustaining Russia's Arctic Cities
Resource Politics, Migration and Climate Change
Edited by Robert W. Orttung

Sustaining Russia's Arctic Cities

Resource Politics, Migration, and Climate Change

Edited by
Robert W. Orttung

berghahn
NEW YORK • OXFORD
www.berghahnbooks.com

Published by
Berghahn Books
www.berghahnbooks.com

© 2017, 2018 Robert W. Orttung
First paperback edition published in 2018

All rights reserved. Except for the quotation of short passages
for the purposes of criticism and review, no part of this book
may be reproduced in any form or by any means, electronic or
mechanical, including photocopying, recording, or any information
storage and retrieval system now known or to be invented,
without written permission of the publisher.

Library of Congress Cataloging-in-Publication Data

Names: Orttung, Robert W., editor.
Title: Sustaining Russia's Arctic cities : resource politics, migration, and climate change / edited by Robert W. Orttung.
Description: New York : Berghahn Books, 2016. | Series: Studies in the circumpolar north ; volume 2 | Includes bibliographical references and index.
Identifiers: LCCN 2016023836 (print) | LCCN 2016041741 (ebook) | ISBN 9781785333156 (hardback : alk. paper) | ISBN 9781785333163 (ebook)
Subjects: LCSH: Cities and towns—Russia, Northern. | Sociology, Urban—Russia, Northern. | Russia, Northern—Population. | Russia, Northern—Environmental conditions. | Social change—Russia, Northern. | Climatic changes—Russia, Northern. | Energy industries—Social aspects—Russia, Northern.
Classification: LCC HT145.R8 S87 2016 (print) | LCC HT145.R8 (ebook) | DDC 307.760947—dc23
LC record available at https://lccn.loc.gov/2016023836

British Library Cataloguing in Publication Data

A catalogue record for this book is available from the British Library

ISBN 978-1-78533-315-6 (hardback)
ISBN 978-1-78533-842-7 (paperback)
ISBN 978-1-78533-316-3 (ebook)

Contents

List of Figures, Maps, and Tables vii

Acknowledgements xi

Preface xii

Chapter 1. Russia's Arctic Cities: Recent Evolution and Drivers of Change   1
  Colin Reisser

Section I. Decision Making

Chapter 2. The Arctic in Moscow   25
  Elana Wilson Rowe

Chapter 3. The Anna Karenina Principle: How to Diversify Monocities   42
  Nadezhda Yu. Zamyatina and Alexander N. Pelyasov

Section II. Migration Trends in Russian Arctic Cities

Chapter 4. Boom and Bust: Population Change in Russia's Arctic Cities   67
  Timothy Heleniak

Chapter 5. Assessing Social Sustainability: Immigration to Russia's Arctic Cities   88
  Marlene Laruelle

Chapter 6. The Russian North Connected: The Role of Long-Distance Commute Work for Regional Integration   112
  Gertrude Saxinger, Elena Nuykina, and Elisabeth Öfner

Section III. Climate Change

Chapter 7. Cities of the Russian North in the Context of Climate Change 141
Oleg Anisimov and Vasily Kokorev

Chapter 8. Access to Arctic Urban Areas in Flux: Opportunities and Uncertainties in Transport and Development 175
Scott R. Stephenson

Chapter 9. Russian Arctic Cities through the Prism of Permafrost 201
Dmitry Streletskiy and Nikolay Shiklomanov

Chapter 10. Urban Vulnerability to Climate Change in the Russian Arctic 221
Jessica K. Graybill

Chapter 11. Conclusion: Drivers of Change 240
Robert W. Orttung

Index 246

Figures, Maps, and Tables

Figures

Figure 1.1. General Typology of Russian Arctic Cities   12

Figure 3.1. Expenditures Supporting Small Business Enterprises in YNAO Cities   44

Figure 3.2. New Determinants for a City's Economic-Geographical Situation   45

Figure 3.3. Expenditures on Culture in Gubkinsky and Other YNAO Cities, 2011   55

Figure 3.4. Small Business and Museum Development in YNAO Cities   59

Figure 3.5. Impact of the Economic-Geographical Situation (EGS)   60

Figure 4.1. Size of the Urban Population in the Russian North, 1989 and 2010   75

Figure 4.2. Population Size in Selected Arctic and Northern Cities, 1897 to 2010   83

Figure 5.1. Net Intra-*oblast* Migration to Tyumen, Khanty-Mansisk, and Salekhard from Other Cities in 2012   101

Figure 5.2. Nationalities of Russian Citizens in Some Arctic Regions   103

viii | *Figures, Maps, and Tables*

Figure 7.1. Temperature Variations at Individual Stations and Regional-Mean MAAT Smoothed with an 11-year Running Filter   147

Figure 7.2. Regional-Mean MAAT Projections from Individual CMIP5 Models, Ensemble of all 36 Models, and Optimal Ensemble of Models with the Best Regional Skills   155

Figure 7.3. Projected Regional Changes in the Cumulative Amounts of Precipitation in the Period October–May (A), and in the Snowfall Period with Temperatures Below 1°C (B)   161

Figure 7.4. Projected Changes of Thawing Degree-Days (ddT), °C d   162

Figure 7.5. Projected Regional-Mean Changes in the Characteristics of the Heating Regime   164

Figure 7.6. Projected Air Temperature Changes for the Spring Break-up Period (May) Relative to the 1961–1990 Norm   166

Figure 8.1. Total Ship-Accessible Marine Area in the Russian Maritime Arctic   182

Figure 10.1. A Conceptual Model of Urban Vulnerability   232

Maps

Map 1.1. Russian Cities   6

Map 4.1. Northern and Arctic Cities   79

Map 6.1. Main Routes of Long-Distance Commute Workers in the Northern Urals and Western Siberia   118

Map 7.1. Location of Weather Stations, the Main Population Centers in the Russian North, and the Southern Permafrost Boundary; Map Partitioned into Regions with Coherent Temperature Changes in the Period 1970–2010   146

Map 7.2. Projected 2010–2015 Changes in the Characteristics of Ice-Jam Floods Relative to the Baseline Period 1946–1977   165

Map 9.1. Study Area and Location of Urban Settlements in Russia Relative to Permafrost Distribution   202

Map 9.2. Changes in Mean Annual Air Temperature between the 1970s and 2000s — 211

Map 9.3. Changes in Foundation Bearing Capacity between the 1970s and 2000s — 213

Tables

Table 3.1. Institutional Aspects of Gubkinsky's and Muravlenko's EGS — 46

Table 3.2. Local Community Commissions and Consultative Councils — 50

Table 3.3. Gubkinsky and Muravlenko Museum Indicators — 54

Table 3.4. Unique Themes in the Cities' Online Forums — 58

Table 4.1. Population Trends in the Russian North, 1989–2013 — 71

Table 4.2. Population of the Oblast Centers in the Russian North, 1989–2010 — 77

Table 4.3. Population Change in Russian North Cities Over 50,000, 1897–2010 — 81

Table 5.1. Foreign-Born Russian Citizens in the 2002 and 2010 Censuses — 96

Table 5.2. Foreign Citizens on the Territory of the Russian Federation, May 2014 — 97

Table 5.3. Foreign Citizens Engaged in Legal Labor Activities in the Russian Federation — 98

Table 7.1. Linear Trend Coefficient for Regional-Mean Air Temperature and Precipitation for the 1976–2010 Period — 148

Table 7.2. Differences between the Modeled MAAT Trends and Observations in the 1976–2005 Period for Selected Regions in the Russian North — 152

Table 7.3. Russian North City Characteristics — 156

Table 7.4. Projected Changes in the Regional-Mean Climate Characteristics — 171

Table 8.1. Annually Averaged Changes in Marine and Terrestrial Accessibility by Midcentury (2045–2059) vs. baseline (2000–2014)   181

Table 8.2. Average and Standard Deviation of Navigation Season Length   183

Acknowledgements

The authors are grateful to the George Washington University Elliott School of International Affairs SOAR Project Initiation Fund for providing the seed money that helped launch this project. Additionally, we would like to thank the National Science Foundation for funding our Research Coordination Network-Science Engineering and Education for Sustainability: Building a Research Network for Promoting Arctic Urban Sustainability in Russia (award number 1231294). Finally, a big thank you to Carrie Schaffner for help with the index.

Preface

Russia's urban areas and associated industrial sites are the location of some of the most intense interactions between man and nature on the planet. Accordingly, a focus on Arctic cities is crucial to understanding Russia's potential for sustainable development. The future of these cities will have an impact on what happens to the rest of the country as well as the global environment.

This book represents the first step in what we see as a multistage study of Arctic urban sustainability in Russia. The purpose of the current volume is to lay the foundation for understanding urban sustainability in the north by examining the broad trends that provide the basis for understanding what kind of challenges Russia's city managers face in planning the future development of their cities.

Specifically, the chapters collected here provide an overview of the three key drivers that influence sustainability in Russia's northern cities:

- policy-making processes,
- resource development and its related labor force requirements, and
- climate change.

These drivers define the nature of the human-environment interaction that takes place in the Arctic. Understanding what these challenges are sets the stage for examining the urban policies city managers adopt to address them and beginning to figure out what works and what does not.

Appreciating the forces shaping Russia's Arctic cities and their ability to achieve sustainability requires bringing together a multidisciplinary team of scholars working in both the natural and social sciences. The chapters gathered here include the work of natural scientists working on issues of climate change and infrastructure resilience as well as social scientists who are more interested in policy making, demographic trends, and understanding the way that Russians conceptualize issues like climate change and sustainability. The book features contributions by climatologists, geographers, political scientists, sociologists, and anthropologists.

The chapters are written from the perspectives of Russians living in Russia, Russians living in the West, and Westerners who are interested in Russia's sustainable development and have made it a part of their research agenda. This multinational team seeks to understand Russian urban Arctic development on its own terms as well as in cross-national perspective.

While the following chapters include some discussion of city-level processes and urban management practices, that is not the focus of this volume. Our next book, research for which is already underway, will look at specific Russian Arctic cities in cross-urban perspective, explaining how the different urban areas are addressing the challenges discussed here. However, in order to get down to the city level of analysis, it is necessary to understand the context within which the cities are located, and that is largely what we do here. With that proviso, readers specifically interested in Russian cities will find lots of new analysis and information in these pages, as each author has attempted to link back their larger themes to Russian Arctic cities.

Defining Terms

Before proceeding to the analysis, it is necessary to explain what we mean by the terms *Arctic* and *sustainability*. "Arctic" has a wide range of definitions across varying bodies of literature, and the interdisciplinary nature of this book necessitates developing a broadly inclusive definition. For the purposes of this project, "Arctic" encompasses territory in Russia whose key features are extreme temperatures for much of the year, pronounced isolation from the urban and industrial cores of the country, lack of permanent transport infrastructure, and the presence of permafrost. Accordingly, warmer northern places such as St. Petersburg are excluded, while much colder and more remote places, such as Yakutsk and Irkutsk, are included. While this loose definition is meant to incorporate the range of chapters in the book and avoid getting bogged down in definitional issues, some specific chapters provide their own narrower definition of Arctic in order to be more precise in their analysis (see, especially, Timothy Heleniak's chapter).

Sustainability as a concept often eludes precise definition, with many scholars and policy makers conceptualizing it differently, often leading to a cacophony of "sustainababble" (Engelman 2013).[1] For purposes of intellectual clarity, we rely on the original definition of the concept from "Our Common Future," the report of the World Commission on Environment and Development chaired by Gro Harlem Brundtland, as meeting "the needs of the present without compromising the ability of future generations to meet their own needs" (World Commission on Environment and Development 1987). Long-term economic stability, a lower environmental impact from natural resource extraction, and a social sphere that meets the needs and aspirations of Arctic residents comprise the central elements of sustainability for the cities of the Russian North (Orttung and Reisser 2014).

For resource-producing centers, the sustainability discourse presents an additional problem, as the production of nonrenewable resources is inherently unsustainable. The resource curse literature has demonstrated a strong connection between resource extraction and economic volatility (Ahrend 2005; Åslund 2005; Gel'man 2010; Ross 2012). Moreover, the Arctic is deeply affected by climate change, with rising temperatures reducing sea ice and making it possible to extract resources, while at the same time thawing permafrost is undermining the foundation for the very infrastructure needed to bring Arctic oil and gas to market (Anisimov and Reneva 2006; Streletskiy, Shiklomanov, and Nelson 2012). Because energy extraction industries are the economic lynchpin of many northern regions (and indeed of much of the Russian economy as a whole), the concept of sustainability in Russia can be adapted to the Brundtland Commission's idea of socioeconomic sustainability. Within this definition, resource extraction activities can be considered sustainable as a way to increase the overall level of social and economic development as long as the environmental impact does not undermine the ability of future generations to meet their own needs (Langhelle, Blindheim, and Øygarden 2008). Because natural resource extraction is such a crucial component of the economy of the Russian North, any conception of sustainability must, at least under the current economic circumstances, integrate continued resource production with the social and environmental needs of northern peoples. At the same time, Russia should be working to overcome the Soviet legacy (Gaddy and Ickes 2013; Hill and Gaddy 2003) while also making provisions for a time when oil and gas are no longer the centerpiece of the economy since neither the Soviet legacy nor the reliance on hydrocarbons is sustainable over the long term.

Plan for the Book

The book proceeds in the following way. Colin Reisser's introductory chapter "Russia's Arctic Cities: Recent Evolution and Drivers of Change" provides the overall context for the more specific chapters that follow. First, it traces the history of urban settlement and development in the Russian North from the imperial period to the present. Second, the chapter provides a simple typology of the Russian Arctic cities to emphasize the differences across them. Third, Reisser lays out the three central change drivers—centralized decision making, resource development, and climate change—that are shaping the growth and decline of Russia's northern cities. These drivers provide a consistent framework for the subsequent chapters that assess Russian Arctic policy-making, immigration, labor policy trends, and the impact of climate change on Arctic infrastructure and accessibility.

Following the introduction, Section I focuses on decision making. Elana Wilson Rowe's chapter "The Arctic in Moscow" starts by examining the policy-making process for the Arctic in Russia's capital. While acknowledging the centralized nature of the process, her analysis emphasizes the large number of players involved and the complex manner required to move from decisions to implementation. The leadership has to engage in a number of balancing acts: pursuing profits versus the need to provide subsidies; centralizing power versus allowing regional leaders more autonomy; and focusing on security versus emphasizing economic development. A review of Russia's key governmental strategy documents for the Arctic shows that they all mention sustainable development, but significant questions remain about what activities will follow from these declarations, how effectively environmental interests will be taken into account, and how much funding will be provided in practice. An analysis of policy debates about the Arctic in Moscow shows that the question of "security" versus "cooperation" garnered the most attention, but that there was little focus on questions of financing Arctic development and federal-regional relations. With many groups joining the discussion, the emerging direction was to pursue military security and economic enrichment simultaneously. Notably absent from the Moscow debate, however, are regional governments and indigenous organizations.

There was also little discussion of the impact of climate change in the debates analyzed. However, there was a general consensus that there is no more need to build cities in the Far North and that it is much preferable to rely on shift workers, who fly in and out of the region.

Following the analysis of decision making at the federal level, Nadezhda Yu. Zamyatina and Alexander N. Pelyasov's chapter "The Anna Karenina Principle: How to Diversify Monocities" looks at the experience of two cities in the Yamal-Nenets Autonomous Okrug: Gubkinsky and Muravlenko. Their analysis provides an explanation for why Gubkinsky has done a much better job diversifying its economy than Muravlenko. The answer lies in the economic-geographic location of the two cities in relation to where administrative and corporate decisions are made: Muravlenko is relatively close to the regional subcenter Noyabrsk and therefore became entrenched in that city's elite networks, which limited the development of small business and cultural institutions like museums. Gubkinsky is located farther away and was able to develop more independently. While this argument relies to some extent on the path dependence of the location of the cities, it demonstrates that local policy makers can have a beneficial impact on how their cities evolve.

Section II "Migration Trends in Russian Arctic Cities" opens with an introductory chapter by Timothy Heleniak, "Boom and Bust: Population Change in Russia's Arctic Cities," that lays out the big picture. Heleniak points out that the situation for the Russian Arctic since the collapse of the Soviet Union has been one of overall shrinkage, with declines in most areas and growth focused mainly in the oil- and gas-producing regions. Increasingly, the Arctic populations are leaving the rural areas of the region and moving into the cities. Overall, the situation seems to be stabilizing and the cities that exist now should serve as the basis for future discussions of sustainability. However, even though today's Arctic cities are not shrinking, Heleniak points out that there is considerable churn in their populations, as many Arctic residents leave for more southern cities while others arrive seeking economic opportunities.

Marlene Laruelle's chapter "Assessing Social Sustainability: Emigration to Russia's Arctic Cities" deepens our understanding of these population movements by looking at the social changes taking place in the cities due to the arrival of immigrants from nearby rural areas, the North Caucasus, and abroad, especially Azerbaijan and Central Asia. Laruelle is primarily interested in issues of equity, diversity, social cohesion, quality of life, and governance. Migrants have become important sources of labor in the extractive industries, construction, public sector waste management, and the broader service sector (restaurants, domestic caretakers). They have integrated into Russia's Arctic cities to varying degrees, with some competing for good jobs and others occupying abandoned professional niches by taking work no one else wants. The chapter notes that Russia has struggled to integrate the migrants into society and protect their rights. The arrival of non-Russian emigrants is creating ethnic districts, altering existing social hierarchies, and transforming the ways that city managers, corporate employers, and citizens interact.

The final chapter in this section addresses the impact of shift workers on the Russian Arctic: "The Russian North Connected: The Role of Long-Distance Commute Work for Regional Integration" by Gertrude Saxinger, Elena Nuykina, and Elisabeth Öfner. After providing a portrait of the shift workers, the authors point out that they have not replaced permanent settlements in the Russian Arctic, but have been operating in the region even as the cities continue to develop. The flow of fly in/fly out workers from southern regions like Bashkortostan has provided links to Arctic cities such as Vorkuta and Novyy Urengoy that have benefitted both the sending and receiving ends in social and economic terms at the micro level. At the same time, the authors draw attention to the deteriorating conditions for workers in the North. In particular, they fear that current practices will bring down living standards in northern energy hubs even as they raise such standards in poor and rural areas in Russia's regions.

The third section of the book looks at the issue of climate change. In "Cities of the Russian North in the Context of Climate Change," physical geographers Oleg Anisimov and Vasily Kokorev provide an overview of how climate change is affecting Russia's North and what these changes mean for the cities located there. While there is a warming trend overall, the authors emphasize that climate change impact will vary from city to city, often depending on local conditions. The consequences of climate change affect the cities through negative impacts, such as permafrost thawing and river flooding, and positive trends, such as a reduced need for heating and decreasing snow depth and duration. This broad overview provides the basis for more detailed studies of transportation connections and permafrost stability.

Scott R. Stephenson examines "Access to Arctic Urban Areas in Flux: Opportunities and Uncertainties in Transport and Development." He shows that warming will have contradictory implications for transportation in the Arctic: increasing ice melt in the Arctic seas will improve access for Northern Sea Route shipping and increase opportunities for off-shore drilling, while warmer temperatures will reduce the number of days per year in which it is possible to use the ice roads, which are crucial for making deliveries to many northern cities. The impact of the increased shipping is extremely uncertain for Russia's Arctic coastal cities. Stephenson argues that ultimately "national development priorities, rather than existing climate conditions, are the main driver for investments in the Arctic and its related urban development, though the state of the sea ice and rivers will have a major impact on the focus of the investment."

In their chapter "Russian Arctic Cities through the Prism of Permafrost," Dmitry Streletskiy and Nikolay Shiklomanov examine the impact of warming on permafrost, providing a quantitative analysis of its continuing ability to support buildings in Arctic cities and infrastructure, such as pipelines, associated with the energy sector. They note that climate change's impact can be intensified in areas of concentrated human activity. In particular, they point out that "observed climate warming has the potential to decrease the bearing capacity of permafrost foundations built in the 1970s," a finding that could have significant implications for the future of Arctic cities.

To conclude this section and tie the discussion of climate change back to where the book began in terms of policy making, Jessica K. Graybill examines the issue of "Urban Vulnerability to Climate Change in the Russian Arctic." She builds on the three preceding technical chapters by affirming a point made by Anisimov and Kokorev that historically Russia has played a leading role in developing climate science. However, the constant flux of agencies and rules at the federal level in contemporary Russia has hindered regional and local responses, including those by corporations and nongovernmental organizations. She then addresses the way that local and indigenous peoples perceive climate change to explain the kinds of policies they are (and are not) willing to adopt in response. While the vast majority of Russians are aware of climate change, they do not see it as a serious personal threat. In particular, Western understandings of climate change have failed to gain much traction. Finally, Graybill builds a methodology for assessing the vulnerability of Russian cities in the Arctic. Among the most vulnerable are the monoindustrial towns, small port cities, and indigenous settlements. Ultimately, social networks can help to address these vulnerabilities. However, Graybill concludes that currently "there is almost no discussion regarding climate change as a basis for decision and policy making regionally, let alone for Russian Arctic cities."

In the conclusion, Robert Orttung synthesizes the findings of the combined chapters to summarize what this volume tells us about

Russia's decision making, resource-driven migration patterns, and climate change, and their impacts on urban sustainability. He then identifies areas for further research that can build on the findings presented here.

Of course, these summaries do not convey the conceptual and empirical richness of each of the chapters in the volume. Hopefully, the reader will be interested enough to follow up with several or all of them.

Notes

1. In order to preserve intellectual consistency with previously published works, the following two definitional paragraphs are drawn verbatim from Orttung 2015.

References

Ahrend, Rudiger. 2005. "Can Russia Break the 'Resource Curse'?" *Eurasian Geography and Economics* 46(8): 584–609.
Anisimov, O.A. and S. Reneva. 2006. "Permafrost and Changing Climate: The Russian Perspective," *Ambio* 35: 169–75.
Åslund, Anders. 2005. "Russian Resources: Curse or Rents?" *Eurasian Geography and Economics* 46(8): 610–17.
Engelman, Robert. 2013. "Beyond Sustainababble," in Worldwatch Institute (ed.), *Is Sustainability Still Possible?* Washington, DC: Island Press.
Gaddy, Clifford G. and Barry W. Ickes. 2013. *Bear Traps on Russia's Road to Modernization*. London: Routledge.
Gel'man, Vladimir. 2010. "Introduction: Resource Curse and Post-Soviet Eurasia," in Vladimir Gel'man and Otar Marganiya (eds), *Resource Curse and Post-Soviet Eurasia: Oil, Gas, and Modernization*. Plymouth: Lexington Books.
Hill, Fiona and Clifford G. Gaddy. 2003. *The Siberian Curse: How Communist Planners Left Russia Out in the Cold*. Washington, DC: Brookings Institution Press.
Langhelle, Oluf, Bjørn-Tore Blindheim and Olaug Øygarden. 2008. "Framing Oil and Gas in the Arctic from a Sustainable Development Perspective," in Aslaug Mikkelsen and Oluf Langhelle (eds), *Arctic Oil and Gas: Sustainability at Risk?* London: Routledge.
Orttung, Robert W. 2015. "Promoting Sustainability in Russia's Arctic: Integrating Local, Regional, Federal, and Corporate Interests," in Susanne Oxenstierna (ed.), *The Economics of the Russian Politicised Economy: Institutional Challenges Ahead*. London: Routledge.
Orttung, Robert W. and Colin Reisser. 2014. "Urban Sustainability in Russia's Arctic: Lessons from a Recent Conference and Areas for Further Investigations," *Polar Geography* 37(3): 193–214.
Ross, Michael L. 2012. *The Oil Curse: How Petroleum Wealth Shapes the Development of Nations*. Princeton: Princeton University Press.

Streletskiy, D.A., N.I. Shiklomanov and F.E. Nelson. 2012. "Permafrost, Infrastructure and Climate Change: A GIS-Based Landscape Approach," *Arctic, Antarctic and Alpine Research* 44(3): 368–80.

World Commission on Environment and Development. 1987. *Our Common Future*. Oxford: Oxford University Press.

CHAPTER ONE

Russia's Arctic Cities
Recent Evolution and Drivers of Change

Colin Reisser

Siberia and the Far North figure heavily in Russia's social, political, and economic development during the last five centuries. From the beginnings of Russia's expansion into Siberia in the sixteenth century through the present, the vast expanses of land to the north represented a strategic and economic reserve to rulers and citizens alike. While these reaches of Russia have always loomed large in the national consciousness, their remoteness, harsh climate, and inaccessibility posed huge obstacles to effectively settling and exploiting them. The advent of new technologies and ideologies brought new waves of settlement and development to the region over time, and cities sprouted in the Russian Arctic on a scale unprecedented for a region of such remote geography and harsh climate.

Unlike in the Arctic and sub-Arctic regions of other countries, the Russian Far North is highly urbanized, containing 72 percent of the circumpolar Arctic population (Rasmussen 2011). While the largest cities in the far northern reaches of Alaska, Canada, and Greenland have maximum populations in the range of 10,000, Russia has multiple cities with more than 100,000 citizens. Despite the growing public focus on the Arctic, the large urban centers of the Russian Far North have rarely been a topic for discussion or analysis.

The urbanization of the Russian Far North spans three distinct "waves" of settlement, from the early imperial exploration, expansion of forced labor under Stalin, and finally to the later Soviet development of energy and mining outposts. To understand the current dynamics, opportunities, and challenges defining the region today, examining the history of the region's development is an important first step.

This chapter provides an overall introduction to the book. We begin with a concise overview of the history of Russian Arctic development from the imperial era to the present. The second section offers a general typology of Russian Arctic cities to emphasize the differences among them. The third section lists the three drivers of change affecting the future of Russia's Arctic cities: centralized decision making, energy development, and climate change.

Arctic Urban Evolution from the Tsars to Post-Soviet Russia

Tsarist Era

Indigenous peoples have long inhabited the Russian Arctic, and while many still live in the region today, the majority of the population is Slavic in origin, as a result of migration from the south. The first round of Siberian settlement followed Russia's conquest of the region in the sixteenth and seventeenth centuries, and was of relatively low intensity. The crown established frontier outposts to defend the newly gained territory, and the growth of these military settlements provided a limited amount of economic opportunity, primarily in the form of fur trapping (Bobrick 1992). The total population of Siberia remained relatively small until around 1800, when political, social, and technological factors converged to make large-scale settlement possible.

By 1800, the Russian Empire effectively had claimed most of the territory of Siberia, yet it remained sparsely populated and largely unproductive (Bobrick 1992). Agriculture was significantly more difficult to pursue in Siberia than in the warmer and more fertile lands of European Russia. By the time that the *Chernozem* (the belt of fertile soil across Southern Russia) was completely divided up between various subjects of the empire, the excess population of Russia needed new land to cultivate, and Siberia presented a potent opportunity. The southern reaches of the territory were increasingly accessible after a number of surveys had been completed in the early 1800s, and their climate was warm enough to support small-scale agriculture (Hill and Gaddy 2003). As the government increased its focus on settling Siberia, a growing number of immigrants began flowing in. Older settlements, such as Krasnoyarsk, entered a phase of rapid growth, and new cities, such as Novosibirsk, were founded.

While a number of factors drove increased settlement in Siberia, the largest single influence in the pre-Soviet era undoubtedly was the construction of the Trans-Siberian Railroad, which began in 1891. Previously, networks of rivers and portages had been the only way to travel across the vast distances of Siberia, with voyages taking weeks or even months to complete. Further, large-scale trade was impossible due to the complicated and lengthy transit system. The imperial authorities were cognizant of Siberia's transport problems, and construction of the Trans-Siberian Railroad was pursued to both ensure reliable transportation between European and Asian Russia as well as to give the military greatly increased deployment ability across the vulnerable eastern frontier. The Trans-Siberian's construction greatly facilitated access to the burgeoning Siberian cities, and it became the chief transport artery for Asiatic Russia long before the entire network was completed in 1916. During the course of the nineteenth century, the population of Siberia grew by between four and five million inhabitants, with the majority of new immigrants arriving during the largest phase of railway construction between 1891 and 1900 (Gilbert 2002). While Southern Siberia grew rapidly during the pre-Soviet period, the Far North remained almost unpopulated until political and technological developments finally opened the frontier in the early twentieth century.

Stalinist/Gulag Era

While the nineteenth century saw a vast expansion of development in Siberia and the Arctic relative to previous levels, the early Soviet period ushered in an era of growth that dwarfed all previous phases. The Russian Revolution of 1917 proclaimed the victory of the urban proletariat in a country that was overwhelmingly rural and agrarian, with rural peasants comprising over 80 percent of the population (Hill and Gaddy 2003: 66). In order to facilitate the transition of the agrarian economy to an industrialized socialist model, the new leadership undertook rapid urbanization, resulting in the total urban population of the USSR growing over 6 percent per year between the 1920s and 1941 (Goskomstat 1998: 32–33). Most of the initial growth occurred in the already-important cities of the European USSR, such as Moscow, Leningrad, and Odessa, as industrialization had yet to proceed in any concentrated fashion east of the Urals. In addition to the socialist mandate that cities must be the core of the state, much of the state-driven urbanization in the USSR aimed to reverse the population shrinkage in major cities caused by the conflict and privations of the Russian Civil War.

In the early years of the Soviet Union, more widespread settlement in Siberia became a major priority for government planners. Taking advantage of technological improvements such as ice-breaking ships, the government's desire to populate the eastern regions of the country began to seem much more realistic. Despite such advances, however, officials had little success encouraging greater migration rates, even with the provision of higher salaries and concerted ideological campaigns meant to bring greater numbers of people to help exploit Siberia's vast resources.

After the lack of success with initial drives to encourage the movement of European Soviet citizens to Siberia and the Far North, the Gulag assumed the central role in the industrialization of Asiatic Russia. While the forced labor system in the USSR started out primarily as a way to either reform or dispose of political prisoners and violent criminals, repeated demands for more workers in developing regions forced authorities to rethink the provision of labor within the country. Regional authorities began demanding access to the growing pool of convict labor, and repressions increased quickly in scale to meet the new labor demands, especially after the initiation of several grand projects such as the White Sea–Baltic Canal (Applebaum 2003: 73). It cannot be said definitively whether labor demands were primarily responsible for the growth of the camps (Stalin's motives being inscrutable at best), but the escalating need for convict laborers, combined with the increasing political repressions, caused prisoner numbers to swell to over a million by 1934, having been just over 20,000 five years earlier (Applebaum 2003: 91).

Prisoner labor was seen as an almost infinite reserve for the exploitation of Siberia's vast resources, and the state's total power over the economy allowed far greater numbers of workers to be sent to Siberia and the Far North than the market would have dictated (Hill and Gaddy 2003). The industrialization of the USSR, as guided by the five-year plan system devised by Stalin, demanded massive exploitation of natural resources to fuel the building of factories and infrastructure, and the need to supply these resources resulted in major expansions of forced labor colonies. In an effort to satisfy the Engels Dictum, which asserted that socialist development must be spread equally across a country's territory, much of the new industry in the Soviet Union was planned to be built in previously undeveloped and remote regions (Koropeckyj 1967). As the most untouched region of the country at the time, the Far North became particularly attractive for expansion. New technologies made access to the remote (but resource-rich) northern regions more viable, and the ideology of the state suggested that concerted effort in the development of communism would be able to overcome the harsh conditions that had previously prevented any large-scale settlement of the region in the past (Koropeckyj 1967: 235).

During the expansion of the Gulag, many cities were founded primarily as labor camps in the Far North, most with a focus on exploiting a single natural resource exclusively through prison labor. Cities such as Vorkuta (1932), Norilsk (1935), Igarka (1929), and Magadan (1929) began functioning during the early years of the Gulag, while older Siberian and northern cities received large influxes of prison labor as well (see Map 1.1). The large semiurban camps were augmented by many smaller satellite camps for logging and other smaller-scale industries meant to support the operation of the central locations. The purges of the mid-1930s greatly increased the available number of inmates to supply the ever-growing labor demands of the camp system, such that by the height of the purges in 1938, there were 1,800,000 Gulag prisoners in the USSR, with a further 1,000,000 in exile (Applebaum 2003: 113).

The rapid industrialization of the USSR during the 1930s required huge inputs of raw materials, and the abundance of such materials in the Far North meant that some of the largest camps developed in the region. The Pechora coal basin near Vorkuta began production to augment the already proven resources of the Donbass in Ukraine; the mining and smelting plants of Norilsk supplied crucial metals that were badly needed to produce steel and advanced machinery; and still other camps were primarily developed as earners of foreign convertible currency, such as the gold mines of Kolyma in Eastern Siberia and the lumber exporting ports of Igarka and Arkhangelsk (Shabad 1969).

The boom in Siberian population due to prison camps had resulted in unprecedented growth in the region, especially in places far from the Trans-Siberian Railroad, where the vast majority of the pre-1929 population had lived. Remote cities in Siberia experienced immense growth, with many more than doubling their populations due to prisoner influxes (Applebaum 2003: 113). The beginning of war with Germany in 1941 provided further stimulus to the already rapid growth rate, as the swift German advance forced a huge share of Soviet industry to be relocated far away from the front lines, which generally meant placing factories in Siberia. While much of the industry in Siberia had been redundant with European regions for reasons of strategic duplication, the occupation of mines and factories in Europe by the enemy meant that numerous Siberian enterprises became the sole

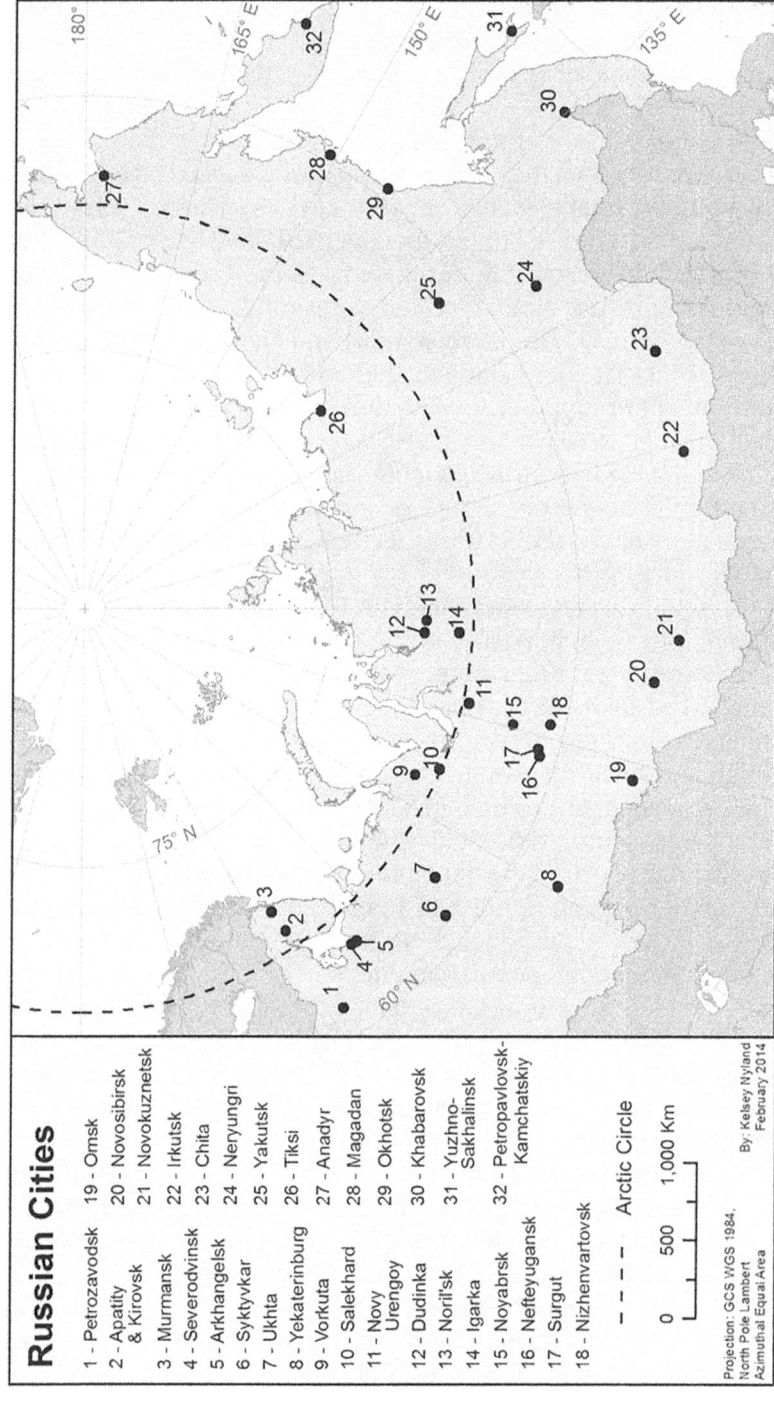

Map 1.1. | *Russian Cities*

provider of their respective materials. In addition, many arms factories were located in the major cities along the Trans-Siberian Railroad (for easy transport access), while the more remote Arctic mines and plants could be used to supply these primary factories by seasonal river supply, or in the case of more rare and valuable materials, by air (Adams 1983).

Cities like Vorkuta and Norilsk received their largest boosts after the original primary supply centers for coal and nickel, which had been near the German front, were shut down, and thus saw their equipment evacuated by rail and barge to Siberia and set up in the secondary locations. In this fashion, previously smaller-scale factories and plants grew significantly in size in the matter of a few months. After the war's end, many of the industries that had been moved to Siberia as a result of hostilities were simply left where they were. The evacuations of people and equipment boosted the development of Siberian regions in a major way. Krasnoyarsk Kray alone experienced population growth of 75 percent between 1940 and 1945 (Shabad and Mote 1977). The wider geographic distribution of industrial centers fit neatly into the ideological framework of the Engels Dictum, and the strategic lessons learned in World War II demonstrated that further industrial development in remote regions would provide greater survivability in the case of another European land war.

Late Communist Era

Stalin's death in 1953 heralded the beginning a new era for urbanization and industry in the Far North, as forced labor fell out of use by the end of the 1950s. In order to adapt to the new economic conditions, northern cities would have to make significant and difficult adjustments. When prisoners in the northern camps were freed, a great deal of them chose to immediately leave for their original homes, and an abrupt population crisis hit the industries of former Gulag cities (Prociuk 1967). In general, the USSR was in a difficult situation regarding the economy: rebuilding the western part of the country after the devastation of the war took enormous resources, and the now rapidly changing demographics of labor forced a complete restructuring of industrial production.

In response to these new conditions, a dual solution arose in the actions of the Soviet government: labor-intensive industries were located in the European USSR or near extant rail transit networks, while electricity- or raw materials–intensive industries, such as mining and smelting, would take place in Siberia, near plentiful sources of materials and electricity that could be generated from hydroelectric dams, coal, and natural gas (Shabad and Mote 1977). To a certain extent, it appears that these changes in location policy constituted an abandonment of the early efforts to ensure equal distribution and Stalinist economics. While Soviet planners seemingly no longer intended to bring development in Siberia to the same level as European Russia, the imperative to distribute industry across the national territory still caused planners to focus a disproportionate level of investment on the remote and inaccessible regions of Siberia and the Far North up until the collapse of the USSR (Hill and Gaddy 2003). Nonetheless, this trend toward resource-based economies in Siberia, instead of the prioritization of production favored by the USSR as a whole, foreshadowed Russia's eventual post-Soviet transition to a deindustrialized economy.

The Manpower Problem and the Northern Shipment

The initial development period following World War II in the Soviet Union had been primarily focused on rebuilding after the devastation caused by four years of war, and as this rebuilding phase came to a close, the country focused on expanding its industrial base. In the postwar Soviet political mindset, Siberia represented the perfect place to expand industry. In ideological terms, location policy and conquering the wilderness to build advanced industry helped the Communist Party achieve its political goals. Military strategy emphasized the benefits of placing industry with military value far from Russia's borders. And, in practical terms, the vast expanses of Siberia contained enormous reserves of every conceivable raw material, from oil to diamonds.

By the 1960s, Siberia and the Far North had become permanent focal points for new projects and investments; the decade saw the discoveries of the first major oil, natural gas, metal, and mineral resources in decades, and new technological methods allowed for the scale of production to be increased far beyond what would have been possible in previous years. The largest single push for developmental expansion came in 1968 with the discovery of the Samotlor oil field in the Khanty-Mansiy Autonomous Okrug in Western Siberia, the first of several "supergiant" oil and gas fields to begin production in the region (Wilson 1987). Large gas fields, like Medvezhye and Yamburg, were discovered further north in the Yamal-Nenets Autonomous Okrug in the 1970s, and these regions of Western Siberia quickly became the largest destinations for investments in the USSR (Shabad

1987). The feverish expansion of oil and gas activities sparked the construction of entirely new cities in Siberia, in what can be seen as the "third wave" of economic and demographic expansion in the region, as distinguished from early imperial settlement and the forced labor expansions under Stalin. Cities such as Novyy Urengoy were built from the ground up, and older ones, such as Surgut, experienced massive increases in population and investment. By the 1980s, Siberia held by far the largest energy and resource reserves in the USSR, with over 80 percent of oil reserves, 90 percent of natural gas, and 90 percent of coal (Wilson 1987). Clearly, the natural resource–based economy of the Soviet Union depended on Siberia as its almost singular resource base.

In order to support the concerted urbanization of Northern Siberia and the Arctic from the 1960s onwards, the government exerted huge efforts through the "Northern Shipment" process (Heleniak 2001). The Northern Shipment effectively consisted of two separate components: subsidies and pay raises granted to individual workers to encourage them to move and remain in the north, and large subsidies for transportation and basic economic resources for northern settlements.

Labor shortages had been a critical problem in the USSR since the introduction of five-year plans in 1928, before large-scale settlement of the Siberian Far North began (Connoly 1987). After the abandonment of forced labor between 1953 and 1957, populations in many northern cities began to decline precipitously, as freed workers moved away en masse. Planners identified the growing gap between labor needed for economic plans and the continued population drain from Siberia as the single largest obstacle to continued development, a phenomenon that came to be known as the "Manpower Problem" in Siberia. According to S.G. Prociuk, the author who coined the term, the environmental difficulties paled in comparison to labor issues in the Far North, noting that "whatever the natural difficulties at the present stage of developing Soviet Siberia may be, they seem to be less than those of transplanting skilled and unskilled manpower to those regions for operating and maintaining new installations (Prociuk 1967: 192). In Krasnoyarsk Kray alone, Prociuk estimated that no more than 12 percent of newly recruited workers even finished their contracts before returning to their original places of residence (Prociuk 1967). The 1.4 million people who moved to Siberia between 1956 and 1960 were offset as between 1.4 and 2 million people left the region during the same period (Prociuk 1967).

In order to reverse these severe declines, the Soviet government devised the "Northern Benefits," an improved system of monetary and social incentives to lure workers to the north and encourage them to remain there for longer periods (Heleniak 1999). Wage augmentation had already been introduced before in certain Gulag camps, both with free and nonfree labor, but the wages offered from the 1970s onwards were unprecedented in their scale in the Soviet Union (Borodkin and Ertz 2003). These programs included not only wage increases, but also extra leave time, paid vacation, and other inducements that were uncommon or completely absent in other regions (Connoly 1987). Additionally, underfulfillment of work norms could not be used as grounds for curtailing bonuses in the Far North, which was one of the most attractive advantages offered (Armstrong 1960).

Costs of living in the Soviet Union were theoretically supposed to be relatively equal across all territories due to the price controls mandated by Soviet economic policy, so the wage differential programs, which offered as much as a 100 percent pay increase over similar jobs elsewhere, would have been one of the few ways to earn substantially higher actual wages (Heleniak 2001). Increased migration levels allowed many major cities in Siberia and the Far North to grow significantly, but even with the expanded system of benefits, luring workers to the north in quantities sufficient to fulfill all economic plans was not nearly as successful as the authorities had hoped (Connoly 1987). For most, the system was less attractive in reality than it seemed on paper; Hill and Gaddy estimate that Siberian costs of living were in reality closer to 35–50 percent higher than those of European Russia, and that average wage differentials amounted to only a 15–20 percent increase, meaning that actual wages in Siberia would have been lower than those in the rest of the country (Hill and Gaddy 2003). Such estimates aside, the urban population of the Far North grew steadily through the mid-1980s, and government programs and incentives no doubt played an important, if not defining, role.

In addition to the Northern Benefits to lure workers to the Far North, the authorities established a logistical and financial system to connect and supply the often remote industrial centers of the region. Due to the enormous geographical separation of Far Northern cities, they were difficult and expensive to supply with necessary basic commodities, such as food and consumer goods. Many cities were only accessible by ice roads or aircraft, so the total costs of provisioning the remotest regions could be enormous. In order to improve living conditions and slightly diversify the work force away from the raw materials sector, some of the larger Siberian cities began to produce limited quantities of basic goods, though usually not in amounts sufficient to satisfy demand (Armstrong 1960). Consequently, the government had to provide massive subsidies for the provision of goods to remote settlements and cities of the north, which consumed enormous financial resources. In addition to subsidizing food and fuel, the "Northern Shipment" even subsidized the costs of basic industrial inputs (coal, primary inputs, etc.). Estimates vary, but as much as 6 percent of the entire Soviet GDP was consumed by this system by the 1980s, demonstrating the incredible inefficiency and enormous total expense of maintaining the large amount of industry that had been set up in the Far North (Heleniak 2001; World Bank 2010).

Post-Soviet Transition

When the USSR disintegrated in 1991, it was clear that the future of many urban centers across the Far North was in jeopardy. The disappearance of state subsidies and political support for expensive and remote industries based in the region meant that neither the excessive populations nor large enterprises could continue in their current forms. Accordingly, much of the 1990s was characterized by deindustrialization, population loss, and political strife as the Far North struggled to reorient itself to a market economy.

Perhaps the most apparent trend during the post-Soviet transition was that of demographic decline. The single-industry cities and towns favored by Soviet planners, known as monocities (*monogorody*), often found themselves without an economic basis once the planned economy disappeared. Consequently, many smaller towns depopulated almost to the point of abandonment, and larger cities declined precipitously. Labor shedding by the industries of the Far North and the ensuing out-migration led to a decline of almost 10 percent of the region's population, 913,000 people, between 1989 and 1997 (Heleniak 1999: 171). Though the pace of depopulation has decreased since 1997, many cities have continued to shrink (for more details, see the chapter by Timothy Heleniak in this volume).

Russia's Diverse Arctic Cities

Despite the numerous changes affecting the Russian Arctic since the collapse of the USSR, the region still remains primarily urban. There are stark differences across the cities and settlements, and they can be sorted according to various criteria. To provide a general overview, the fourfold typology in this section classifies the cities by the level of their economic diversity and their growth status (see Figure 1.1).

	Economic Diversity	
	−	+
Growth Status +	Novy Urengoy: Energy Center	Yakutsk: Growing, Diversified City
Growth Status −	Norilsk: Remote Industrial City	Murmansk: Diversified Transit Hub

Figure 1.1. | *General Typology of Russian Arctic Cities*

Other authors in this volume develop different typologies that are more appropriate to their specific topics (see, in particular, chapters by Timothy Heleniak and Jessica Graybill). While the overall population of the Russian Arctic has been in decline since 1991, some Arctic urban centers have registered growth, making them notable exceptions to this trend. In particular, cities dominated by energy exploitation have fared best in avoiding population decline, as well as a small number of more southerly centers with good infrastructure and transit accessibility, such as Yakutsk. Conversely, cities with a Soviet-era economy based on lower-profit sectors, such as timber or low-value mining (coal), have had the largest population declines. Additionally, the centers that are most inaccessible and remote have had larger declines in general, due to the higher costs of transport and lower level of economic opportunity. The following cities provide illustrative examples of the different types of urban settlements in the Russian Far North—diversified transit hub, energy center, remote industrial city, and growing, diversified city.

Murmansk: Diversified Transit Hub

As the largest Arctic Russian city, Murmansk has been an important center through which much of the region's economic activity passes. Most internal shipping along Russia's Arctic coast either originates from or is received by Murmansk's port, and the city is an important industrial center for processing metals and other mine products. Additionally, Murmansk is an important military center, particularly for the navy, as the Russian Northern Fleet is based nearby. Despite its central importance for the Arctic as a whole, Murmansk's population has shrunk by over 150,000 since 1989, to just over 300,000 in 2010, and has shown no signs of growth under current conditions. While Murmansk is important as a bridge to Arctic centers (the *Atomflot* icebreakers are based there, as is the headquarters of the Northern Sea Route, which allows ships to sail between Europe and Asia across the top of Russia), its permanent land connections and ice-free port insulate it from many of the challenges facing other Arctic centers.

Novyy Urengoy: Energy Center

Novyy Urengoy is perhaps the most significant example of the energy-dominated "growth poles" throughout the Russian Far North, and it is one of the largest and most prosperous northern centers. While the city only grew moderately in the 1990s, its status as the primary Gazprom hub in Siberia has ensured strong governmental support, investment, and growth in recent years. The expansion of energy exploration in the Yamal-Nenets region, largely based out of existing energy centers like Novyy Urengoy, Surgut, and Nadym, as well as the continued dominance of the energy industry in Russia's economy, will likely mean that such centers continue to be foci of development and economic activity.

Norilsk: Remote Industrial City

As the second largest Arctic city, Norilsk has been an important regional center for decades. Norilsk typifies the remote industrial centers of the Arctic, as it has no permanent transit connections, and must rely on aircraft and icebreaking ships to maintain itself. Despite the profitability of the nickel and platinum mining that dominates the city's economy, the population declined markedly in the two decades following the Soviet collapse. A reduction in Norilsk Nickel's workforce has driven much of this shrinkage, as well as a general out-migration from northern cities by more economically mobile citizens. While large industrial centers such as Norilsk are unlikely to disappear completely, continued depopulation is a desirable outcome for the enormous mining concern based there and the federal government so that the high costs of maintaining remote centers can be reduced.

Yakutsk: Growing, Diversified City

Yakutsk represents a small group of diversified cities in the Arctic that are growing. These are traditional regional centers that provide administrative services for the areas around them. They benefit from nearby resource production, diamonds in the case of Yaktusk or gas in the case of Salekhard. As Timothy Heleniak's chapter points out, Yakutsk has grown extensively based on continuing in-migration from outlying areas, including by indigenous peoples.

As the above examples demonstrate, the progress of economic decline and redevelopment during Russia's transition from central planning to market mechanisms has been uneven and highly dependent on local circumstances. The energy industry remains the major driver of investment and development in most urban centers, and those locales that are further from energy exploration and reliable permanent transport infrastructure continue to face economic depression and the resultant demographic decline. A major challenge facing cities and towns of the Russian Far North is to find ways to survive through developing a degree of urban sustainability in such varying conditions, with differing levels of government investment and widely divergent economic prospects.

Drivers of Change in the Russian Far North

Given the historical legacy and current diversity of Russia's Arctic cities, what forces are likely to determine future developments in these settlements? The complex interaction among human and environmental pressures to effect rapid change within the Arctic or other environments has been increasingly described through the agency of "drivers of change" (e.g. Gore 2013; Smith 2011). After the varying booms, busts, and stagnations of the Soviet and post-communist periods in the Russian Far North, several forces are combining to change the future of northern cities. Three distinct forces are at work:

- political centralization,
- energy development, and
- climate change.

All of these forces intersect in the cities of the Russian North. The combined impacts of these three drivers of change are likely to greatly alter the economy, society, and political framework for the cities of the Russian North in the coming decades. At a time when the Russian government has clearly outlined its increased focus on the Arctic and its desire to develop the extractive industries of the region, these factors are certain to become even more crucial.

Politics

First, following the economic and social havoc in the early 1990s as Russia transitioned from a planned economy to the market, President Vladimir Putin's rise to power brought efforts to strengthen central authority in Moscow and impose greater political control over the rest of the country. Because the economic future of Russia in many ways depends on the successful extraction of northern resources, the stakes for increasing governmental control have become even higher. The Putin era has therefore been a struggle between regional interests, business interests, and the federal government, with the federal government playing the dominant role. The Kremlin has sought to increase its authority in northern regions while promoting energy development and the necessary local development, working with Russia's energy companies, regional governments, and foreign investors in the process. The ways in which the political landscape has been altered and continues to evolve are critical to the development and success of northern urban centers, and the ever-present hand of the state is sure to play a critical role for northern cities for the foreseeable future.

Since coming to power, Putin has taken a keen interest in the energy sector and has personally directed its development. Therefore, the federal government has a strong presence in the cities largely run by state-owned energy firms such as Gazprom and Rosneft (Novyy Urengoy, Surgut, etc.). The incredible energy resources of regions such as Yamal Nenets make a tight relationship with the Kremlin inevitable; the *okrug* supplies more than 80 percent of Gazprom's total extractive wealth, while tax revenue from energy companies makes up in excess of 90 percent of the region's budget (Kusznir 2006). With a central government so strongly tied to energy exploration in the Far North, northern cities are likely to see an active government presence in coming decades.

The hand of the Kremlin may often be easy to detect through state companies, yet the central government also has increased its involvement with private companies such as Norilsk Nickel and Novatek, both of which had achieved a relatively high degree of independence during the late 1990s and early 2000s. Although Norilsk Nickel was privatized in the early 1990s, government interventions under Putin since 2003 have increasingly called into question the independence of the firm. In addition to planting government-friendly representatives on Norilsk Nickel's board of directors after the departure of then-director Mikhail Prokhorov in 2008, the company was forced to take out large loans from the state during the concurrent financial crisis, further giving the state power to intervene in the affairs of the company (Humphreys 2011). For companies like Norilsk Nickel that dominate their respective monocities, governmental intervention in ostensibly private corporate affairs is a strong indication that the state will maintain a prominent presence in the economic and political affairs of northern cities.

Beyond state-business relations, Putin's policies since 2000 have directly increased the role of the federal government in regional politics. Regional centers lost their ability to directly elect governors in 2004, when the president took the power to appoint them directly. Although limited gubernatorial elections returned in 2012, the Kremlin retains extensive control over who can become a regional executive. The removal of the longtime governor of the Yamal-Nenets Autonomous Okrug, Yuri Neelov, and his replacement with the less independent Dmitry Kobylkin is a prominent example of the Kremlin's desire to maintain a tighter degree of control over the most economically essential provinces of the country (Kusznir 2006).

Given the reality of political centralization in Russia, Chapter 2 examines the nature of the decision-making process for the Arctic in Moscow. But there are limits to central control, and some Arctic cities have managed to develop outside of its reach. Therefore, Chapter 3 examines how two cities in the Yamal-Nenets Autonomous Okrug have pursued sustainability goals, one in the shadow of Russia's administrative edifice and one more independently. The chapter shows that there are numerous development paths even within a centralized system.

Energy Development

Second, given its economic and political importance, the energy industry is the most immediate driver of change in the Russian Arctic on the ground and its needs have played a determinative role in the shape of Russia's Arctic cities. Although resource development has always been at the heart of life in the north, from furs in past centuries to timber and mining at the zenith of Soviet expansion, the energy industry has been the primary economic enterprise in recent decades. While energy exploration was a major driver of growth in the USSR from the 1960s onwards, the deindustrialization of the Russian economy resulted in the country's ever-greater reliance on basic extraction (Jensen 1983). The energy sector could influence Russian Arctic cities in two ways: through the development of new cities, as was the practice in the past, or by directing the flows of people to, or away from, the different northern cities.

The energy boom in the Russian economy during the first fifteen years of the twenty-first century did not spur the development of new cities, so it differed from the kind of development that took place in the Soviet era. The discovery of large oil and gas resources in Western Siberia in the 1960s transformed the Soviet economy, and energy resources have since become the most valuable export for Russia. As the older fields of the 1960s and 1970s decline in productivity, the industry has gradually moved northwards, increasing its presence in the Arctic and sub-Arctic regions of the country. These changes naturally spurred development in Russian cities.

The Soviet model of industrial expansion into Siberia and the Far North called for the construction of cities to support industry, so centers such as Novyy Urengoy (1975) and Nefteyugansk (1967) were built from the ground up, while others such as Surgut and Salekhard were expanded once oil and gas fields were discovered nearby. These energy cities have been among the few urban areas in the remote regions of Russia to continue to grow after the collapse of the Soviet Union, demonstrating their importance to the energy industry as a whole.

The expansions of the energy industry in the Far North underway today have moved further and further away from the previous urban centers established during the height of Soviet energy exploration, with a special focus on the more remote reaches of the Yamal Peninsula. The new projects in the region have brought large infrastructural investments to assist in the expansion of development, though the towns of the region have yet to expand significantly. The Ob-Bovanenko Railway, which connects the Bovanenkovo gas field to Labytnangi, has been one of the largest investments to date, and the presence of permanent transport infrastructure in this previously unsettled area could allow for a degree of urbanization as the local oil and gas industry continues to expand. Further developments, such as Novatek's approved plan to build a large LNG export facility at Sabetta, could ultimately stimulate new urban settlement.

Despite this rapid energy growth, no large settlements have been constructed since the dissolution of the USSR, and it is unclear whether any further centrally planned urbanization will take place on orders from Moscow. Rather than constructing new cities in previously uninhabited areas, the strategy for the large energy companies such as Gazprom, Rosneft, and Novatek has been to rely on shift workers to supply labor to the increasingly remote fields of the north. These workers typically live in camps that do not require the construction of a full complement of city services.

Accordingly, instead of initiating the construction of new cities in the Russian Arctic, energy development has been the key variable in determining which existing cities are growing and which are shrinking. The main impact of the energy sector is felt through the way it is bringing workers to the Arctic. The emphasis on shift workers has changed the nature of the kind of settlement that is taking place. Moreover, the fact that the energy sector is attracting workers from the Caucasus and Central Asia is changing the social fabric of the northern cities. The chapters in section II of this book will examine these trends in greater detail.

Climate

The third driver of change that will affect the Russian Far North is the prospect of climate change. The climate of the region has been one of the largest influences on virtually every aspect of life for northern cities, and the predicted increase in temperature over the coming decades will force most cities to face daunting challenges, in addition to the possibility of some benefits. Transit accessibility, long one of the most difficult and expensive obstacles to urban development in the Far North, is projected to vary wildly. Sea access along the Northern Sea Route is forecast to increase markedly, while land access is almost certain to degrade at a similar pace. The increasing instability of permafrost will likely have catastrophic implications for the structural integrity of buildings and infrastructure in the region, especially as the quality of much northern infrastructure is already significantly degraded at present.

The interaction between the energy industry and the federal government has been and will certainly continue to be a major force shaping the Russian Far North, yet the future of cities in the region is certain to be driven by the outcomes of climate change. Climate change has already begun to make its presence felt in the Arctic due to its increased speed and strength there in what scientists describe as "Arctic amplification" (Jeffries, Richter-Mege, and Overland 2012). While it is clear that climate change is taking place, the exact ways in which it will influence urban life in the Far North are not as well understood. Certain effects of warming promise to increase transit accessibility and mitigate some of the difficulties of Arctic settlement, yet others have more ominous implications for the future of northern cities.

The most obvious implication of warming for the Arctic is that it will lessen the harshness of winter for the north. By 1983, it was estimated that work stoppages in the Arctic due to cold amassed losses of up to 33 percent of all possible working hours, a staggering drain on productivity for a largely industrial region (Mote 1983). Later analysts have attempted to quantify the direct effects of low temperatures on the Russian economy, with Tatiana Mikhailova estimating that 1.2

percent of the total Russian GDP is drained by the urbanization of Russia's coldest regions (Mikhailova 2007). A rise in temperatures could lessen the frequency of work stoppages due to cold, reduce the energy costs of heating for residents, and simply make the north a more livable place.

While the effects of temperature on the livability of the Arctic may prove to be a significant driving force for change in the future, one of the most immediately apparent climatic factors in the Arctic and Far North is the way in which transit accessibility is changing. The lack of affordable and reliable transit access to the Arctic has been a crucial obstacle to development of the region, with transit and supply subsidies from the Soviet and Russian governments, the so-called Northern Shipment consuming up to 6 percent of the national budget at different points in history (Heleniak 2001). The partial loss of these subsidies after 1992 and the resulting economic and demographic decline is indicative of how crucial the provision of transport is to the Russian Far North.

Sea transit along the Northern Sea Route and the major rivers of Russia leading to it (particularly the Ob, Yenisei, and Lena) was a consistent drain on resources in the USSR and Russia since the Northern Sea Route (NSR) was first used in the early Stalinist expansions of settlement. Even to the present, icebreakers are required for safe and reliable transport, leading costs in cities lacking other transit alternatives to remain consistently high (Hill and Gaddy 2003). The progress of climate change has already reduced sea ice along the NSR significantly during much of the year, leading to increased accessibility, and predictions of warming by 2050 that show that sea-based transit in the Arctic may be greatly facilitated (Stephenson, Smith, and Agnew 2011). The energy and mining industries of the Far North already rely heavily on the western NSR for much of their economic livelihood, but the costs of icebreaking and the seasonality of transport are significant economic drains (Ragner 2000). Whether this increased accessibility will result in a greater degree of urbanization in the Far North is difficult to tell, though reduced sea transit costs will certainly benefit industry in the region.

Tempering the increased sea accessibility of the Far North is the reduction in land access, which is currently largely dependent on ice roads, as much of the region lacks permanent road or rail infrastructure (Hill and Gaddy 2003). Warmer temperatures have begun to reduce the length of time during the year when ice roads can be used for transport in remote areas, and the significantly warmer temperatures predicted by most climate models are certain to exacerbate this effect.

A decline in the ice road season could induce a steady rise in transit costs for many northern settlements, unless mitigated by cheaper and more reliable sea access. Noncoastal settlements are likely to be the hardest hit by this trend, as air transit is prohibitively expensive for larger urban centers. Predicting how the differing changes in access will affect urban centers is difficult, and will be determined by the scale and pace of warming, as well as local geographic variation.

In addition to drastically changing the accessibility of northern cities, climate change promises to severely impact the physical integrity of the cities themselves. For cities constructed on permafrost, as many major cities of the Russian North, such as Norilsk, Yakutsk, and Vorkuta are, warming temperatures promise to increase the pace of permafrost thaw, with grave consequences for physical infrastructure. As the ground warms, the stability of building and road foundations decreases rapidly, leading to building deformation and even outright structural collapse (Mazhitova et al. 2004). The pace of deformation increases with the scale of warming, so the projected increases of the coming decades are ominous for northern cities. For example, Norilsk alone had 250 major deformations or collapses by 2003, most of which have required complete building demolition (Ilichev et al. 2003). An analysis of bearing capacity change in Igarka and Norilsk for 2041–2060 predicts declines of 61.5 percent and 40 percent, respectively, which would result in a catastrophic number of building collapses and deformations (Streletskiy 2012). For the extant larger cities of the Far North, the rapid decline in infrastructural stability may be the most difficult and expensive problem to overcome in the coming decades. Despite the awareness of the progression and risks of climate change within the academy, engagement with the topic has been sparse within the federal government. In the most recent Russian policy document on Arctic development, climate change is acknowledged as a potential factor in the region, though no steps are outlined to mitigate its effects or adjust development accordingly (Pravitelstvo RF 2013).

This chapter has launched the book by presenting the three drivers of change for Russia's Arctic cities. The following chapters will examine the nature of decision making, resource-driven migration flows, and the impacts of climate change on Russia's Arctic cities. The conclusion will tie together the main themes and assess the implications for urban sustainability.

Colin Reisser is Senior Geological Technician at EOG Resources, Denver, Colorado.

References

Adams, Russell B. 1983. "Nickel and Platinum in the Soviet Union," in Robert Jensen (ed), *Soviet Natural Resources in the World Economy.* Chicago and London, University of Chicago Press, pp. 536–555.

Applebaum, Anne. 2003. *Gulag: A History.* New York: Anchor Books.

Armstrong, Terence. 1960. "Mining in the Soviet Arctic," *Polar Record* 10(64): 16–22.

Bobrick, Benson. 1992. *East of the Sun: The Epic Conquest and Tragic History of Siberia.* New York: Poseidon Press.

Borodkin, Leonid, and Simon Ertz. 2003. "Coercion versus Motivation: Forced Labor in Norilsk," in Paul R. Gregory and Valery Lazarev (eds), *The Economics of Forced Labor: The Soviet Gulag.* Stanford, CA: Hoover Institution Press, pp. 75–104.

Connoly, Violet. 1987. "Siberia: Yesterday, Today, and Tomorrow," in Rodger Swearingen (ed.), *Siberia and the Soviet Far East.* Stanford, CA: Hoover Institution Press, pp. 1–39.

Gilbert, Martin. 2002. *The Routledge Atlas of Russian History.* London and New York: Routledge.

Gore, Al. 2013. *The Future: Six Drivers of Global Change.* New York: Random House.

Goskomstat. 1998. *Naseleniye Rossii za 100 let (1897–1997): Statichesky sbornik.* Moscow: Goskomstat.

Heleniak, Timothy. 1999. "Out-Migration and Depopulation of the Russian North during the 1990s," *Post-Soviet Geography and Economics* 40(3): 155–205.

———. 2001. "Migration and Restructuring in Post-Soviet Russia," *Demokratizatsiya* 9(4): 531–49.

Hill, Fiona and Clifford Gaddy. 2003. *The Siberian Curse: How Communist Planners Left Russia Out in the Cold.* Washington, DC: Brookings Institution Press.

Humphreys, David. 2011. "Challenges of Transformation: The Case of Norilsk Nickel," *Resources Policy* 36: 142–48.

Ilichev, V.A., V.V. Vladimirov, A.V. Sadovsky, A.V. Zamaraev, V.I. Grebenets, and N.B. Kutvitskaya. 2003. *Perspektivy razvitiya poseleniy Severa v sovremennikh usloviyakh* [Prospects for the development of settlements of the North in present conditions]. Moscow: SOPS.

Jeffries, M.O., J.A. Richter-Menge, and J.E. Overland, eds. 2012. "Arctic Report Card 2012." National Oceanic and Atmospheric Administration. Retrieved 22 February 2013 from http://www.arctic.noaa.gov/reportcard.

Jensen, Robert. 1983. "Soviet Natural Resources in a Global Context," in Robert Jensen (ed.), *Soviet Natural Resources in the World Economy.* Chicago and London: University of Chicago Press, pp. 3–10.

Koropeckyj, Iwan S. 1967. "The Development of Soviet Location Theory before the Second World War," *Soviet Studies* 19(2): 232–44.

Kusznir, Julia. 2006. "Gazprom's Role in Regional Politics: The Case of the Yamalo-Nenets Autonomous Okrug," *Russian Analytical Digest* 1(6): 10–12.

Mazhitova, Galina, Nanka Karstkarel, Naum Oberman, Vladimir Romanovsky, and Peter Kuhry. 2004. "Permafrost and Infrastructure in the Usa Basin (Northeast European Russia): Possible Impacts of Global Warming," *AMBIO: A Journal of the Human Environment* 33(6): 289–94.

Mikhailova, Tatiana. 2007. "The Cost of Cold: The Legacy of Soviet Location Policy in Russian Energy Consumption, Productivity, and Growth." Unpublished paper. Boston University.

Mote, Victor. 1983. "Environmental Constraints to the Economic Development of Siberia," in R. Jensen, T. Shabad, and A. Wright (eds), *Soviet Natural Resources in the World Economy*. Chicago: University of Chicago Press.

Pravitelstvo RF. 2013. "Strategiia razvitiia Arkticheskoi zony Rossiiskoi Federatsii i obespecheniia natsional'noi bezopasnosti na period do 2020 goda" [Strategy of development of Arctic regions of the Russian Federation and national security in the period to 2020]. Retrieved 8 July 2014 from http://government.ru/news/432.

Prociuk, S.G. 1967. "The Manpower Problem in Siberia," *Soviet Studies* 19(2): 190–210.

Ragner, Claes Lykke. 2000. "Northern Sea Route Cargo Flows and Infrastructure—Present State and Future Potential." Fridtjof Nansen Institute Report 13/2000. Retrieved 2 April 2011 from http://www.fni.no/doc&pdf/FNI-R1300.pdf.

Rasmussen, Rasmus. 2011. "Megatrends." Nordic Council of Ministers. Retrieved 4 March 2012 from http://www.nordregio.se/en/Publications/Publications-2011/Megatrends/.

Shabad, Theodore. 1969. *Basic Industrial Resources of the USSR*. New York and London: Columbia University Press.

Shabad, Theodore. 1987. "Economic Resources," in Alan Wood (ed.), *Siberia: Problems and Prospects for Regional Development*. London, New York, and Sydney: Croom Helm, pp. 62–95.

Shabad, Theodore and Victor Mote. 1977. *Gateway to Siberian Resources (The BAM)*. Washington, DC: Scripta Publishing.

Smith, Laurence. 2011. *The New North: The World in 2050*. London: Profile Books.

Stephenson, Scott R., Laurence C. Smith, and John A. Agnew. 2011. "Divergent Long-Term Trajectories of Human Access to the Arctic," *Nature Climate Change* 1: 156–60.

Streletskiy, Dmitry. 2012. "Projections of Climate and Bearing Capacity in Igarka and Norilsk." Unpublished paper.

Wilson, David. 1987. "The Siberian Oil and Gas Industry," in Alan Wood (ed.), *Siberia: Problems and Prospects for Regional Development*. London, New York, and Sydney: Croom Helm, pp. 96–129.

World Bank. 2010. "Implementation Completion and Results Report, Northern Restructuring Project." World Bank Report. Retrieved 18 October 2011 from http://go.worldbank.org/SVO5AIZLL0.

Section I

Decision Making

CHAPTER TWO

The Arctic in Moscow

Elana Wilson Rowe

Introduction

Russian president Dmitry Medvedev and Norwegian prime minister Jens Stoltenberg proudly announced in Oslo in May 2010 that the two countries had settled their competing claims over a sector of the Barents Sea and had finally agreed to a delimitation line that had been an object of sporadic negotiations for over 40 years. The result of the long-term discussions was that the two countries divided the disputed area of overlapping claims neatly between them. Medvedev spoke optimistically about the accord as a "cooperation agreement" and argued that Arctic countries "need to resolve difficult questions, only in that way can we look to the future" (Latukhina 2010).

While the delimitation agreement was widely reported in the Norwegian press, it hardly made the evening news in Russia. The day after this announcement, then Prime Minister Vladimir Putin was visiting Franz Josef Island and spoke explicitly about Russia's national geopolitical interests in the North, emphasizing border control, economic interests, and the presence of significant military infrastructure (Petrov 2010). The timing of the visit may have been a coincidence, but it also may have been a way of showing a domestic audience that, although Russia had settled in international negotiations for less than it had originally claimed, the defense of national interests in the symbolically and economically important Arctic was still paramount.

Of course, these two episodes reflect the Kremlin leadership's appreciation of their differing audiences—one international and one domestic—and what had the most political currency in each setting.

However, these moments also highlight that the Arctic presents complex opportunities and challenges for Moscow and that policy thinking about the region is by no means set in stone. The extent of this intricacy and potential for evolution is perhaps not surprising when you take into account how vast, geographically varied, and economically important the region is. What is defined as "the High North" and so-called "areas equivalent to the High North" encompass more than 60 percent of Russian territory.[1] Were this region an independent state, it would constitute the world's largest country. Although sparsely populated, with only 8.2 million residents in 2006, the North accounts for 20 percent of Russian GDP and 22 percent of all Russian exports (Gusher 2009).

The region also factors into and is affected by broader changes in Russian security thinking and practices, which brings an additional layer of complexity to Arctic policy making. Russia's annexation of Crimea in the spring of 2014 triggered a low point in Russia-European/North American relations and a sanctions regime against the country, including measures targeted to hinder development of Russia's Arctic energy sector (Conley and Rohloff 2015: 2). International military and security-related cooperation was immediately frozen in the region, although low-level safety/security relations continued, for example, coast guard cooperation (Østhagen and Gastaldo 2015). On the level of diplomacy and cross-border relations on nonsecurity issues, Russia continues to engage as before, demonstrating that the region is seen as one where Russian interests are well served by continued cooperation and seemingly seeking to minimize spillover effects from conflict elsewhere (Wilson Rowe and Blakkisrud 2014; Conley and Rohloff 2015).

While this chapter does not detail international cooperation/diplomacy and security issues in the Arctic, these are important geopolitical features to keep in mind in light of Arctic urban development. For example, a more security-oriented Russian Arctic is likely more "closed" to outside involvement—with external commercial/financial actors hindered from engaging by sanctions or reputational concerns—thereby stalling or further complicating the large-scale natural resource development that has been the historical driver for regional development (see Chapter 1 of this volume). Another angle to consider is whether increased military spending will benefit the development of Arctic infrastructure more broadly also for commercial/social purposes, or whether the Arctic may be a "loser" in light of generally expanded military spending and efforts to secure Russia's position along its now troubled western border.

Many of the chapters in this book give us concrete insight into the challenges and strategies utilized in Arctic urban environments themselves. This chapter attempts to provide a broader view of some of the enduring balancing acts and cross-cutting tensions that mark Russia's approach to Arctic development. Given the centralization of Russia's political system today, the politics and policies of the federal center are essential to understanding the background against which Arctic urban development is unfolding.

Russia and Policy Making

The political environment in which Arctic policy making takes place is a relatively centralized and vertically organized one that merits some attention before we turn to more specifically Arctic policy issues. Russia's main political action unfolds in Moscow, where strong central figures like President Putin play a publicly decisive role in most key political decisions, with the rest of the government (the Duma, various ministries) generally following suit. Nevertheless, even though Russia is far from a functioning democracy, characterizing the country as subject to one-person rule overlooks much of the complexity in Russia's political processes.

Richard Sakwa's (2010) description of Russia as a "dual state" brings to the forefront some of the complexity, tension, and dynamism of political processes in Russia. Sakwa suggests that we think of Russian politics as a struggle between two systems—the formal constitutional order (termed the "normative state") and "a second world of informal relations, factional conflict and para-constitutional political practices," which he calls the "administrative regime" (Sakwa 2010: 185). While the politics of the administrative regime—with its elite factions, grey eminences, and murky public-private relations—has likely rightfully attracted most academic and popular attention, constitutional commitments do sometimes serve as a source of discursive power and the "formal niceties" of the constitution remain an important source of popular legitimacy in domestic politics (Sakwa 2010:18).

Furthermore, policy making in all states—dual or otherwise—is an inherently complex process, particularly in making policy statements into implemented realities. Some policy fields in Russia function in a more pluralistic fashion than one might expect in a state with weak checks and balances (Wilhelmsen and Wilson Rowe 2011). While Putin and other key figures have the deciding voice publicly on many issues, it is logical to expect that there are multiple foci of decision making that allow for the influence of various actors. Such diversity of access and influence points may hold true especially for the "low politics" of social and environmental issues, as opposed to the "high politics" of national security and questions of war and peace.

Arctic policy making—with a wide range of social and environmental issues, implications for economic growth, and security concerns—includes issues that fit into both the high and low ends of the policy making spectrum. This combination lends a dynamism to Arctic policy making, which can be fruitfully approached as both a regionally focused policy field in itself and also as a conglomerate of multiple national-level broader policy fields (economic, security, social, and environmental) with regional aspects.

The specifically regional and also cross-cutting nature of Arctic policy problems in Russia (as in other countries) may have contributed to the many reorganizations and changes in terms of where Arctic issues are handled and coordinated in Moscow. The Ministry of Regional Development had been the longstanding focal point for Arctic issues, including interministerial coordination, international cooperation, and indigenous affairs. This ministry was dissolved as "superfluous" in September 2014, with responsibility for Arctic issues divided amongst the ministries of economic development, justice, finance, and culture (Staalesen 2014). This reorganization took place concurrently with discussions about the need for a single body that could take responsibility for implementation of Arctic policy. The "Arctic Commission" was officially established in February 2015 and headed by the long-serving Deputy Prime Minister Dmitry Rogozin, who is in charge of defense and space issues and is also known for making internationally controversial claims about Russian power and intentions (Pettersen 2015).

While Russia's top politicians may (although rarely do) take varied approaches to Arctic politics in Russian public discourse, it is likely that a range of civil servants and policy entrepreneurs further down Moscow ministerial hierarchies and in other state bodies play an important role in shaping Russia's Arctic politics. Before further examining who participates in Arctic policy today, we will first briefly look at Arctic policy in a historical perspective and then carry out a review of key Arctic policy documents.

Past Practices and Long-Term Balancing Acts

As discussed in Chapter 1, northern resources played an important part in the Soviet planned economy, while opening and developing the North (*osvoenie severa*) filled a corresponding role in Soviet ideology. The Soviet focus on the North positioned the Arctic firmly as a factor in both Russian national identity and conceptions of economic and state security (Blakkisrud 2006). Single-industry towns built up around key Arctic resources, forced labor populations, and military installations all contributed to the Soviet and now Russian Arctic being the most decidedly urbanized of all the circumpolar countries (Arctic Council 2004). This emphasis on the North, however, resulted in Russia inheriting from the Soviet Union an "overpopulated" region ill-suited to the demands and logic of a market economy (see Chapter 4). In today's Russian Arctic, I would argue that there are four main tensions or balancing acts that characterize Russian policy making around development questions.

First, there is a tension between traditions of, and continued need for, large state subsidies of Arctic infrastructure and social services, and the desire to have the Arctic be primarily a source of profit for the entire country. Russian Northern policy during the transitional 1990s could be described as haphazard and focused primarily on emergency measures to respond to economic and social crises in the region. The contours of a more clearly discernible Russian policy on the North emerged under Putin's first two terms (2000–2008). This approach was initially based on principles of market economics with an eye towards ensuring that the North became a profitable part of the Russian state (Blakkisrud 2006). The region's natural resource wealth made income generation a realizable pursuit—to some extent. The Arctic produces about one-tenth of the world's crude oil and a quarter of its gas. Of this output, 80 percent of the oil and 99 percent of the gas come from Russia (AMAP 2007). However, there are limits to a market-driven development of the North that continuously raise the issue of subsidies versus market mechanisms and private investment. This tension is one that some of the strategy documents presented below address directly and is particularly prominent in large infrastructure projects, such as the rejuvenation of ports along the Northern Sea Route or renewal of the icebreaking fleet (Moe 2014).

A second issue is the locus of decision-making power. Putin's recentralization of power from the regions to the federal level contrasted sharply with the widespread decentralization of the 1990s (Blakkisrud and Hønneland 2000). Moscow rather than Magadan or Murmansk now governs this vast territory. At the same time, the cooperation of regional governors is essential for implementing federal policies in far-flung regions of Russia; they are often called upon to publicly front Arctic efforts and, thereby, likely exert some behind-the-scenes influence. The balance of power between regional and federal government remains an important one for understanding the political and budgetary impetuses and constraints in Arctic city- and regional-level government. At the present juncture, however, this relationship does not garner much attention in the Arctic policy documents and media debates reviewed below. Despite the difficulties and inefficiencies that a strong centralization in such a large and diverse country may bring about, the leading role of the federal center in Arctic issues does not seem to be open to question—at least not in Moscow.

A third balancing act—between an "open" and "closed" Arctic—also characterizes the region. Specifically, Russia's evolving relationship to its North entails a tension between the securitization of northern space and the nationalization of northern resources working against more international and market-driven orientations (Wilson Rowe 2009). For example, Soviet Arctic industrial cities—in particular those associated with the military complex—were among the more closed places in the Soviet space, requiring special and closely controlled registration permits even for Soviet citizens. At the same time, the natural resources around which many of these cities were built were subject to global affairs and commodity markets, leaving them vulnerable to the vagaries of international politics and price swings.

Marlene Laruelle, in her comprehensive book on the Russian North, describes a similar tension, coining it as one between a "security first" and a "cooperation/economics first" reading of the region (Laruelle 2013: 7). She argues that the Federal Security Service, the military-industrial complex, and President Putin prioritize security since they see the Arctic as a platform from which Russia can assert its "great power" status. The "cooperation first" approach draws inspiration from an emphasis on economic opportunities and the necessity of garnering investment and gaining access to foreign expertise. Proponents of this approach include the Ministry of Natural Resources, the Ministry of Regional Development, and Prime Minister Medvedev (Laruelle 2013: 7).

Finally, and of particular relevance to this book's emphasis on problems of urban sustainability, a fourth balancing act is between commercial and environmental concerns. The demise of the Soviet Union in 1991 left Russia with serious environmental issues, as the Soviet regime had largely failed to protect the environment from the negative consequences of industrial development (Rowe 2013; Oldfield 2005; Ostergren and Jacques 2002). As historian Lars Rowe points out in his study of the Soviet nickel industry, environmental protection and coping with pollution were left to the same ministries that were responsible for promoting industrial development, and a deeply utilitarian view of nature prevailed (Rowe 2013: 11–17).

This is not to say that Soviet society was devoid of concern for nature. Despite being primarily subservient to industrial concerns, the Soviet regime developed environmental monitoring infrastructure and environmental expertise and practices (Oldfield 2005; Bruno 2011). The most influential and noticeable outlet for Soviet environmental interest was a movement that argued for protecting significant tracts of land from industrial development in the first place—the *zapovednik* system. Such a focus on "pristine" nature was more acceptable to the Soviet leadership, in part because it upheld a division between industrialized areas and wilderness areas (Weiner 1999). Today, while the Russian public is concerned with environmental quality and less willing to "pay the costs of pollution" (Whitefield 2003: 102; Crotty and Hall 2012), these concerns have not been linked to significant action and environmentalists have been relatively weak political actors throughout the post-Soviet period (Henry 2010: 764; Yakblokov 2010).

On a similar note, the term *sustainable development* remains a "convenient rhetorical flag … under which ships of many different kinds can sail" (Adams 1990: 3), in Russia as elsewhere. As discussed in the introduction to this volume, whether oil and gas development (a fundamentally unrenewable resource) can be considered "sustainable development" at all—regardless of how socially and environmentally successful the projects—is a question of relevance to the primarily extractive industry based economy of the North. Nonetheless, the term *sustainable development* (*ustoichivoye razvitie*) is common currency in the Arctic development strategies reviewed below (as it is in most circumpolar strategy documents across the region). However, the actual practice of sustainable development remains contested and variable. To take one example, Canadian and Russian understandings of the concept in an Arctic context were shown to differ considerably in terms of what kinds of activities could possibly be sustainable and what kinds of relationships between stakeholders and involved parties a sustainable development approach to natural resources would imply (Wilson 2007).

Current Policy Documents

The past practices and enduring balancing acts—between state subsidies-private investment, center-periphery, open-closed, industrialization-environmentalism—described in the preceding section form the broader historical and discursive backdrop against which new policies relevant to the Arctic are developed. On the whole, the Arctic is increasingly being handled as an integrated aspect of policy fields (including economic, social, and energy policies) as opposed to an object of particular policy attention (Blakkisrud 2006; Wilson Rowe and Blakkisrud 2014). For example, both the current national security and energy strategy documents emphasize the importance of Arctic oil and gas to the future of the country (Laruelle 2013: 5 and 135) as does the recently released Maritime Doctrine (Russian Federation 2015).

Despite the trend towards handling the Arctic through regular channels, there are documents with a specifically Arctic profile, three of which are reviewed here. These strategies are the 2008 *Fundamentals of State Policy of the Russian Federation in the Arctic in the Period up to 2020 and Beyond* (hereafter *Arctic Strategy*) (Security Council 2008), the *Strategy for Socio-economic Development of Siberia towards 2020* (hereafter *Strategy for Siberia*) (Government of Russia 2010), and the Arctic-specific follow-on from the Siberia strategy titled *Strategy for the Development of the Arctic Zone of the Russian Federation* (hereafter *Arctic Development Strategy*) (Government of Russia 2013).

The Arctic Strategy from September 2008 (Security Council 2008) presents the Arctic first and foremost as a "strategic resource base" to secure socioeconomic development of the country. The document makes clear that Russian policy makers are keenly aware of the value of Arctic resources and see them as a key element in the long-term future of the Russian economy. Other national interests include preserving the Arctic as a zone of peace, protecting the Arctic environment, facilitating rejuvenation of the Northern Sea Route, increasing security (through better border control, an Arctic military brigade, and improved search and rescue arrangements), making a common information space in the Arctic zone, modernizing Arctic transport, and maintaining science and research capacity through cooperation with other states. In many ways, this document was primarily oriented outwards, providing signals about Russia's favorable attitude towards international cooperation and its general vision for the Arctic. The publication of this short document corresponded with the general flurry of Arctic documents that was published by the United States, Canada, Russia, Denmark/Greenland, and Norway following the 2008 Illullissat Declaration. Nonetheless, the strategy names a number of policy areas that are deemed central to Arctic domestic development—for example, by emphasizing the Northern Sea Route. These points are not, however, presented in any detail.

The *Strategy for Socio-Economic Development of Siberia Towards 2020* (Government of Russia 2010), by contrast, points us more forcefully in the direction of domestic development issues and thus merits more attention in this volume. It is an extensive and detailed document, presenting challenges, general visions for the region, and place-specific prioritizations. There are a number of statements in the text that give the reader a general sense of the political rhetoric surrounding Arctic development and cities at present more generally, which are not officially covered in the strategy.

The overall aim of the strategy is to ensure sustainable development of the region, which involves achieving a level of welfare comparable to that in central Russia, and to meet national security/foreign policy objectives. More specific regional priorities include intensification of geological mapping and exploration, development of new sources of natural resources, resurrection and development of the Northern Sea Route along Russia's Arctic coastline, protection of the environment, and preservation of the culture of indigenous peoples. A great deal of space is devoted to a variety of means to achieving these ends including public and private investment, modernization of social systems (education, health), emphasis on value-added processing of natural resources, and development of transport infrastructure. International relations are also seen as important in realizing the strategy's aims, in particular establishing an export infrastructure to the rapidly expanding economies of the Asia Pacific.

The policy document lists challenges to realizing the policy's goals, including the region's reliance on natural resources (and, by extension, vulnerability to global price changes), capital flight, lack of economic diversity, deteriorating infrastructure (the words "restore" and "resurrect" are used frequently), low wages, and "social depressiveness." Nonetheless, the policy document attempts to recast some of Siberia's geographical challenges (vastness, remoteness) as positives—forwarding the idea of Siberia as a "transport bridge" between the countries of Western Europe, North America, and East Asia. Existing and expanded railway connections, a revived Northern Sea Route with updated port facilities, use of the region's rivers, and even ambitious plans to create a rail link across the Bering Sea and into Alaska are named as regional possibilities to achieve this more geopolitically prominent bridge position.

The policy summaries and recommendations are further divided into three geographical concepts—an Arctic, a northern, and a southern "belt." Of particular relevance to the studies in this book are the Arctic and northern zones. The Arctic zone includes the northern parts of Krasnoyarsk Kray, while the northern belt includes the northern parts of Omsk, Tomsk districts, central parts of Krasnoyarsk Krai, the north of Irkutsk district, and the northern parts of the Buryat Republic.

The economic future of the Arctic belt is clearly stated to be natural resource development, namely, the further development of the metallurgical and petroleum sectors (including offshore development). Concomitant aims are the protection of indigenous peoples' livelihoods, modernization of transport (road, air) infrastructure, and the successful establishment of a culture of shift-work and mobile villages. A clear concern in the document is the potential for unwanted growth of new permanent settlements and stationary populations around natural resource projects. This concern receives as much attention and space in the document as concern for environmental impacts of natural resource development.

It seems that the challenge the new Russian state faced in managing its "overpopulated" North has made a lasting impression. Both government bodies and private companies (who often have a large number of responsibilities for the settlements that grow up around extraction projects) likely have worked to ensure that a shift-work approach to Arctic belt development received such a prominent place in this strategic document. Here it is interesting to speculate that Russian policy makers are attempting, to the extent possible, to decouple economic exploitation of the Arctic from demographic growth in the region. By contrast, this concern about permanent settlements is much less pronounced when it comes to the more temperate "northern belt" and the range of development options presented are more diverse (forest products, high-tech, production of consumer and industrial goods).

The *Strategy for Arctic Development* can be read as a kind of zone-specific follow-on policy to the *Strategy for Siberia*, although this strategy also covers the European Arctic that was excluded from the Siberian strategy. Regardless, the form and formulations are much like the broader Siberian strategy and will thus only be reviewed briefly here. The policy opens by rehearsing the standard statement on Russia's overall aims for the Arctic—the realization of national interests and the pursuit of sustainable development. This preamble is followed by a quite detailed and sobering recitation of the challenges facing the Arctic region relating to declining population, challenging climatic conditions, distance from major centers of trade, low-quality social services, low capacity in search and rescue, an aging icebreaker fleet, and many more issues. The list of policy aims is equally broad—covering aspirations for social issues, economic development (tourism and natural resource development in particular), maritime safety/development of the Northern Sea Route, and sovereignty concerns. The protection of indigenous lifeways—as well as a stated aim of ensuring that indigenous youth are educated to be able to take part in "modern" careers relating to Arctic development—occupies a more prominent position in this document than it does in the Siberia strategy. The "minimization of environmental damage" is also mentioned briefly, with greater attention devoted to questions of environmental monitoring.

Interestingly, the question of how all this development will be financed is raised repeatedly throughout. In this document, the long-term balancing act outlined above between state subsidies and private investment with profit as an aim is directly addressed in several places. The model the strategy seems to be proposing as the ideal to be pursued when possible is public-private partnerships. Important to note as well is that repeated emphasis is made on funding scientific research, like meteorology, geological surveys, and environmental monitoring. However, urban-level and regional-level adaptation to the ongoing climate change effects detailed elsewhere in this volume receives only scant mention, as do the politics and environmental issues raised by climate change as a global problem more broadly. This lack of detail is perhaps unsurprising given Russia's somewhat ambivalent relationship to international climate science and international climate change politics (Wilson Rowe 2013).

As with any of the often quite ambitious Russian strategy documents, a critical question meriting further research is the extent to which these development aspirations garner budget allocations for implementation. The next section attempts to, at least obliquely, address this issue of implementation by presenting the wider Arctic debate in which these strategic documents are generated and received by policy actors. One could argue that the aspects of the policy documents that are most robustly linked to dominant political discourses and that enjoy the support of major policy actors are most likely to be pursued.

Arctic Policy Debates and Actors

In order to take the temperature of Arctic policy debates in Moscow and highlight which actors are involved in these debates, I draw upon a study published in 2014 that analyzed Russian Arctic discourse through a systematic review of 323 articles on Arctic themes in the state-owned newspaper *Rossiiskaia gazeta* (hereafter RG) (see Wilson

Rowe and Blakkisrud 2014 for full description of method and results). To broadly characterize these results, one could argue that of the four "balancing acts" or policy tensions described above, "security first" versus "cooperation" garnered the most attention. Less attention was directed to questions of financing Arctic development, balancing environmental and industrial concerns, and regional-federal relations.

In terms of the "open or closed" balancing act, we noted that a great deal of the coverage in a cooperative tone was generated by international collaborative activities ranging from high-level ministerial meetings in the Arctic Council to new programs in the University of the Arctic cooperative educational network (Wilson Rowe and Blakkisrud 2014). Statements about Arctic cooperation were along the lines of Minister of Foreign Affairs Sergei Lavrov's response to a question about "the war over Arctic resources": "Truly this is the battle that never started ... we do not share these worrisome prognoses relating to violent conflict of interests in the Arctic" (Shestakov 2008). More competition-oriented approaches are certainly to be found, often coming from academics and security actors. For example, the head of the Security Council represented Russia as being legally minded and peaceful in the Arctic, but stated that the United States was acting competitively by carrying out a northern military build-up and that Norwegian research vessels in the North were actually carrying out espionage (Chichkin 2009).

Overall, however, media coverage representing the Arctic as a zone for cooperation rather than conflict has grown steadily (Wilson Rowe and Blakkisrud 2014). Officials tended to take a position between these perhaps overdrawn extremes of all cooperation or all competition. They took a more conciliatory approach and placed great emphasis on cooperation, while also underlining Russia's commitment to defending its interests in the North. To take one example, at a United Russia party meeting in Yekaterinburg, Putin landed somewhere in the middle between competition and cooperation: "I would like to emphasize that Russia certainly will expand its presence in the Arctic. We are open to dialogue with our foreign partners, with all neighbors in the Arctic region, but, of course, we will defend our own geopolitical interests firmly and consistently" (RG 2011).

Furthermore, the policy problems drawing Russian actors' attention northwards are quite diverse and not solely related to security or natural resources alone. Shipping, research activity, climate change, energy, and broad official statements about the state of the Arctic all received some attention—and there was an overwhelming emphasis on the Russian territorial North rather than international issues or

Arctic waters (Wilson Rowe and Blakkisrud 2014). This emphasis on domestic issues is reflected not only in the amount of coverage but also in the statements made by the key actors themselves. For example, presenting the Arctic Strategy in 2008, Medvedev underlined that the major goal for Russia is to address the "undeveloped economic infrastructure and unsolved social problems ... hindering northern development" (Il'in 2009). He also pointed to the economic importance of the Arctic to the Russian economy and argued that "our first and main task is to include the Arctic into the resource base of Russia in the 21st century" (in RG 2008).

In tandem with an increased attention to the growing diversity of policy issues relating to the Arctic, the range of actors involved in speaking about Arctic politics in RG expanded (Wilson Rowe and Blakkisrud 2014). While the MFA and the Presidential Administration were the dominant voices in 2008 and 2009, by 2010 the debate around the Arctic had spread to other sectors in Moscow and a wide range of ministries and state agencies had come to intervene in the Arctic policy debate. The newcomers included actors like the military and the Federal Security Service, the Ministry of Regional Development (which had particular responsibility for coordinating Arctic policy domestically until its dissolution in 2014), as well as a broader range of governmental ministries working on their specific portfolios (for example, health and transport).

Concluding Thoughts

The growing number of actors participating in Arctic policy making as well as the myriad policy concerns discussed in the pages of RG suggest that the Russian authorities now have an expanded understanding of how sovereignty and national interest can be pursued. Historically, military might but also scientific exploration, mapping, and permanent settlements have been essential to demonstrating state sovereignty over distant Arctic territories (Bravo and Sorlin 2002). In this light, Russia's pursuit of state interests (including sovereignty) is becoming less of a question of the "open/closed" or "security first/ cooperation and economic interests first" dynamic described above but rather an effort to achieve both at the same time in a mutually reinforcing fashion. There do not seem to be any individuals or groups of individuals that lobby for a more closed or "securitized" Arctic that might stand in the way of the economic development goals outlined in the strategy documents reviewed above. At the same time, the broader changes in the security environment following Russia's annexation of Crimea create a situation in which this dual approach may be less tenable or more easily challenged domestically.

The larger group of actors engaged in Arctic policy debates since 2009 is also in keeping with the broader effort to include the Arctic in mainstream policy making rather than treating it as a special object of politics. It also suggests that the Arctic may be seen as an interesting object for policy making—one with a bright economic future and a certain political glamour—that consequently attracts the attention of various authorities in Moscow. They do not want to be left out in the cold, so to speak. The question remains, however, whether government and private actors at both the federal and local level are able to coordinate in such a way so as to make good the ambitious policy documents produced.

Although the connection between participation in media discourse about the Arctic in one official newspaper and actual involvement or participation in relevant policy making is a parallel that should not be drawn too directly, it is also interesting to note whose voices were *not* heard in the pages of *Rossiiskaia Gazeta*. Representatives of regional governments and indigenous organizations, to take two examples, were largely absent. While this is perhaps not surprising to students of Russian politics, given Russia's centralized traditions for government and shaky democratic practices, this is a significant difference from other Arctic states that have at least a stronger rhetorical commitment to hearing the voices of Arctic residents. A similar absence is a concern for the potential impacts of climate change on the region.

Finally, the policy documents and debates reviewed above shed some light on how the Arctic is being represented in policy circles in Moscow. Traces of the past practices and enduring balancing acts (between state subsidies–private investment, center-periphery, open-closed, industrialization-environmentalism) identified in the opening section are noticeable in all of the policy documents, with perhaps the questions of finance garnering the most attention and issues of center-periphery relations the least. Throughout, we see that the Arctic's natural resources are conceptualized as an important treasure chest of future economic development for the whole country and Siberia recast in the role of a "bridge" between Europe and Asia in a positive geopolitical spin on the region's vastness. We also see that while indigenous peoples and protection of their livelihoods are mentioned in all policy documents, the overwhelming emphasis is on shift-work and the need to actively avoid the growth of new settlements in the region. All this would serve to reduce the costs and responsibilities of the Russian state and Russian companies active in the region. While the importance of delivering social services to existing populations is also emphasized, it seems clear that a populated Arctic is no longer an aim in and of itself or to demonstrate sovereignty.

Elana Wilson Rowe is Senior Research Fellow at the Norwegian Institute of International Affairs (NUPI).

Notes

1. The Russian terms are *"Krainii Sever"* and *"mestnosti priravnennye k raionam Krainego Severa."* For the full list of territories included in this definition, see http://www.lawrussia.ru/texts/legal_689/doc689a264x760.htm (retrieved 30 April 2012).

References

Adams, William. 1990. *Green Development: Environment and Development in the Third World.* London and New York: Routledge.
AMAP. 2007. *Oil and Gas Assessment 2007.* Retrieved 30 April 2012 from http://www.amap.no/oga.
Arctic Council. 2004. *Arctic Human Development Report.* Akureyri: Stefansson Arctic Institute.
Blakkisrud, Helge. 2006. "What's to Be Done with the North?" in H. Blakkisrud and G. Hønneland (eds.), *Tackling Space: Federal Politics and the Russian North.* Lanham, MD: University Press of America, pp. 25–52.
Blakkisrud, Helge and Geir Hønneland. 2000. "Center-Periphery Relations in Russia's European North," *Polar Geography* 24(1): 27–56.
Bravo, Michael and Sverkar Sörlin (editors). 2002. *Narrating the Arctic: A Cultural History of Nordic Scientific Practices.* Canton, MA: Science History Publications.
Bruno, A.R. 2011. "Making Nature Modern: Economic Transformation and the Environment in the Soviet North." PhD Dissertation in History, University of Illinois.
Chichkin, A. 2009. "Polyus pochti vzyat [The pole is almost taken]," *Rossiiskaia gazeta,* 12 May.
Crotty, J. and S.M. Hall. 2012. "Environmental Awareness and Sustainable Development in the Russian Federation," *Sustainable Development* 22(5): 311–20, DOI: 10.1002/sd.1542.
Conley, Heather and Caroline Rohloff. 2015. *The New Ice Curtain: Russia's Strategic Reach to the Arctic.* Washington, DC: CSIS.
Government of Russia. 2010. *Strategy for Socio-Economic Development of Siberia Towards 2020.* Retrieved from http://www.rg.ru/2010/11/20/sibir-site-dok.html.
Government of Russia. 2013. Strategy for the Development of the Arctic Zone of the Russian Federation. http://government.ru/news/432.

Gusher, A. 2009. "Arktika—zona strategicheskikh interesov Rossii [Arctic—A zone of Russia's Strategic Interests]. Available at <http://www.cisvmeste.ru/show.html?ac=newsitem&cid=2348>, accessed 30 April 2012.

Henry, L.A. 2010. "Between Transnationalism and State Power: The Development of Russia's Post-Soviet Environmental Movement," *Environmental Politics* 19(5): 756–81.

Il'in, A. 2009. "Arktike opredelyaet granitsy [The Arctic determines borders]," *Rossiiskaia gazeta*, 18 September.

Laruelle, Marlene. 2013. *Russia's Arctic Strategies and the Future of the Far North*. New York: M.E. Sharpe.

Latukhina, Kira. 2010. "More popolam: Rossiya i Norvegia dogovorilis' o demarkatzii granitz [The polar sea: Russia and Norway agreed on the demarcation of the border]." *Rossiiskaia gazeta*, 28 April. http://www.rg.ru/2010/04/28/medvedev.html.

Moe, Arild. 2014. "The Northern Sea Route: Smooth Sailing Ahead?" *Strategic Analysis* 38(6): 784–802.

Oldfield, J.D. 2005. *Russian Nature: Exploring the Environmental Consequences of Societal Change*. Aldershot: Ashgate.

Ostergren, D. and P. Jacques. 2002. "A Political Economy of Russian Nature Conservation Policy: Why Scientists Have Taken a Back Seat," *Global Environmental Politics* 2(4): 102–24.

Østhagen, A. and V. Gastaldo. 2015. "Coast Guard Co-operation in a Changing Arctic." Munk Gordon Arctic Security Program Report.

Petrov, V. 2010. "Rossiya namerena provesti v Arktike 'general'nuya uborku' [Russia intends to conduct a 'general cleaning' in the Arctic]," *Rossiiskaia gazeta*, 24 September.

Pettersen, Trude. 2015. "Moscow Kicks Regional Development out of Ministry." 6 February. Retrieved 17 December 2015 from http://barentsobserver.com/en/security/2015/02/controversial-politician-head-artic-commission-06-02.

Rossiiskaia gazeta (RG). 2008. "Dmitrii Medvedev: Arktka dolzhna stat' resursnoi bazoi Rossii [Dmitrii Medvedev: The Arctic should become a resource base for Russia]," *Rossiiskaia gazeta*, 17 September.

RG. 2011. "Vladimir Putin: Rossiya rasshirit svoe prisutstvie v Arktike [Vladimir Putin: Russia is expanding its presence in the Arctic]," *Rossiiskaia gazeta*, 30 June.

Russian Federation. 2015. Morskaya Doktrina Rossiskoi Federatsii (The Maritime Doctrine of the Russian Federation). Retrieved 17 December 2015 from http://static.kremlin.ru/media/events/files/ru/uAFi5nvux2twaqjftS5yrIZUVTJan77L.pdf.

Rowe, Lars. 2013. "Pechenga Nikel: Soviet Industry, Russian Pollution and the Outside World," PhD dissertation. University of Oslo.

Sakwa, R. (2010). *The Crisis of Russian Democracy: The Dual State, Factionalism and the Medvedev Succession*. [Online]. Cambridge: Cambridge University Press. Available at: http://www.cambridge.org/gb/academic/subjects/politics-international-relations/russian-and-east-european-government-politics-and-policy/crisis-russian-democracy-dual-state-factionalism-and-medvedev-succession?format=PB.

Security Council. 2008. Osnovy gosudarstvennoi politiki RF v Arktike na period do 2020 i dal'neishuyu perspektivu [The fundamentals of state policy of the Russian Federation in the Arctic in the period up to 2020 and beyond]. Retrieved 5 January 2012 from http://www.scrf.gov.ru/documents/98.html.

Shestakov, E. 2008 "Rossiya dogovorilas' podelit' Akrtiku po-druzheski [Russia agreed to divide the Arctic in a friendly manner]," *Rossiiskaia gazeta*, 29 May.

Staalesen, Atle. 2014. "Moscow Kicks Regional Development out of Ministry." 11 September. Retrieved 17 December 2015 from http://barentsobserver.com/en/politics/2014/09/moscow-kicks-regional-development-out-ministry-11-09.

Weiner, D.R. 1999 *A Little Corner of Freedom: Russian Nature Protection from Stalin to Gorbachev*. Berkeley: University of California Press.

Whitefield, S. 2003. "Russian Mass Attitudes Towards the Environment, 1993–2001," *Post Soviet Affairs* 19(2): 95–113.

Wilhelmsen, Julie and Elana Wilson Rowe. 2011. *Russia's Encounter with Globalization*. Houndmills, UK: Palgrave.

Wilson, E. 2007. Arctic Unity, Arctic Difference: Mapping the Reach of Northern Discourses. *Polar Record* 43(225): 125–133.

Wilson Rowe, E. 2009. "Afterword: The Intersection of Northern and National Policies," in E. Wilson Rowe (ed.), *Russia and the North*. Ottawa: University of Ottawa Press.

Wilson Rowe, E. 2013. *Russian Climate Politics: When Science Meets Policy*. Houndmills, UK: Palgrave.

Wilson Rowe, E. and H. Blakkisrud. 2014. "A New Kind of Arctic Power? Russia's Policy Discourses and Diplomatic Practices in the Circumpolar North," *Geopolitics* 19(1): 66–85.

Yablokov, A. 2010. "The Environment and Politics in Russia," *Russian Analytical Digest* 79: 2–4.

CHAPTER THREE

The Anna Karenina Principle
How to Diversify Monocities

Nadezhda Yu. Zamyatina and
Alexander N. Pelyasov

Lev Tolstoy's recipe for a happy family is well known: it assumes the confluence of many factors and therefore "all happy families are like each other, while each unhappy family is unhappy in its own way." In each case of unhappiness, there is a unique cause. In a contemporary marriage, the problem could be different views on gender roles, finances, conflicts over how to raise the children, relations with the relatives and neighbors, or a host of other factors.

We apply the same principle to local development. The diversification of "unhappy" monocities—one of the most complicated regional development problems facing contemporary Russia—cannot be achieved by a one-size-fits-all radical solution (Animitsa and Novikova 2009; Kuznetsova and Lyubovnii 2004; Zubarevich 2012). It is especially important to note that the solution does not depend directly on financial resources: otherwise, both the rich and poor would pay. For the happy diversification of the city, there should be a good, complementary fit among all the elements of the local system: the state–property–society. This is the "black box" of regional development. To open it, it is necessary to take into account the economic-geographical situation, which includes, according to contemporary understandings, institutional factors. Doing so makes it possible to stimulate the creativity of the local society, particularly its ability to inspire and nurture innovations, and helps facilitate its evolution into a "learning society"—the most important resource of contemporary economic development at the local level, regardless of specialization.

The difficulty in the mutual interaction of these elements can be illustrated through the comparative analysis of the social-economic development of two neighboring monocities, Muravlenko and Gubkinsky. At first glance, these cities are twins: both are located in the southern part of the Yamal-Nenets Autonomous Okrug (YNAO); they were both founded at the same time (1984 and 1986 respectively); both specialize in extracting hydrocarbons; and their population is similar (25,800 and 33,500 on 1 January 2012). The cities have comparable budget revenues (3.9 billion rubles and 3.3 billion rubles in 2011) mainly supplied by transfers from the YNAO budget. In short, neither city can be considered poor. In both cities, the amount of oil extracted by the main enterprise has fallen in recent decades due to the exhaustion of the nearby deposits.

Only the outcome for the two cities turned out to be different. Gubkinsky looks significantly better than Muravlenko: small business is growing, local firms are making new products, and the city provides more social services for the population and a more comfortable life overall. The company Kirill, one of Gubkinsky's small business enterprises, is the largest producer of dairy products in the region. From 2006 to 2011, small business tax in Gubkinsky formed about 5 percent of city budget revenue, about twice the amount in Muravlenko. Gubkinsky boasts more restaurants and cafes than its neighbor and benefits from more hair dressers, barbers, and shops per 1,000 residents.

The small business of Gubkinsky is distinguished not only by its quantity, but by its quality as a generator of innovation. The majority of registered patents in the city come from its small businesses. For example, in 2011 the Gubkinsky small cosmetology business Aphrodite introduced new Cellulab ultrasound cavitation technology to reduce fat and cellulite and laser technology to provide a younger look for skin. In the sphere of ecology and biotechnology, small business carried out an investment project to create a modern production facility to produce a wide variety of high-quality biodegradable packaging, including polyethylene boxes and bottles.

Entrepreneurship in Gubkinsky started practically at the same time that the socioeconomic development trajectories of the two cities began to diverge. Even at the early stages, the difference was institutionalized with the formation of the Gubkinsky Council of Entrepreneurs, which was founded in 1994.[1] Muravlenko's business community did not receive a similar institution until much later—only in 2002.[2] Gubkinsky opened its first business incubator in 2003, its second incubator in 2007, and third incubator in 2012 for stimulating local production. Muravlenko opened its first incubator only in 2011. Entrepreneurship is one of the most important generators of news in the life of Gubkinsky. By contrast, in Muravlenko, the majority of reporting focuses on events in the social sphere and crime.

Of course, the extensive development of small business—which is extremely sensitive to local institutional conditions—was not achieved in a vacuum. In Gubkinsky small business enterprises have received unprecedented support from the local government, making it an outlier in YNAO (see Figure 3.1).

It would seem that everything is simple: where the authorities had supported small business, it grew.[3] It is important, however, to look deeper. Why, despite the comparable budgets and equivalent natural and economic situations, did one city provide so much greater support to small business? Why is Gubkinsky producing better results? Why did Yamal's largest milk factory not appear in Muravlenko? Why, finally, and this is the most important question, does the development of small business in Gubkinsky correlate with a series of other special features of the local system of authority, property, and society, including a more democratic system of local government, generous financing for the cultural sphere, and the relatively positive mood of local young people?

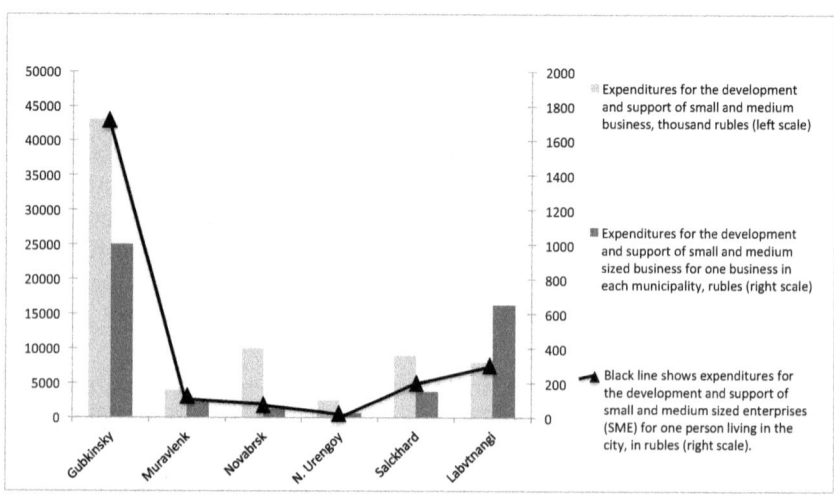

Figure 3.1. | *Expenditures Supporting Small Business Enterprises in YNAO Cities*
Source: City Administrations

The search for answers to these questions according to Russia's geographic traditions usually starts with an analysis of the economic-geographical location of the cities. In this case, location is the key to the answer—and our understanding of economic-geographical location has to be reexamined from a contemporary viewpoint.

A Large Institutional Gap

The key factor generating the difference between the two cities has been the geographical distance between the cities and the subregional center of Noyabrsk, which has a population of 109,000. Muravlenko is 120 km from Noyabrsk (2 hours by road), while Gubkinsky is approximately 240 km (3.5 hours drive). This small geographical distance translated into absolutely different institutional conditions: Muravlenko became an institutional periphery of Noyabrsk, while Gubkinsky was a much more independent subcenter.

Here we must explain how we have expanded the traditional understanding of the concept of the economic-geographical situation (EGS). Traditionally, in examining the EGS an observer takes into account a city's sources of raw materials and energy, road accessibility, and sales market. Recognizing the role of institutional factors for economic development requires expanding the list of conditions that have economic significance. Today it is necessary to take into account the location of the city in relation to the centers of power where decisions are made: administrative centers and headquarters of large corporations, which are particularly important for monocities (See Figure 3.2).

Traditional EGS factors are largely inert and any changes are connected with large financial expenditures (the construction of new

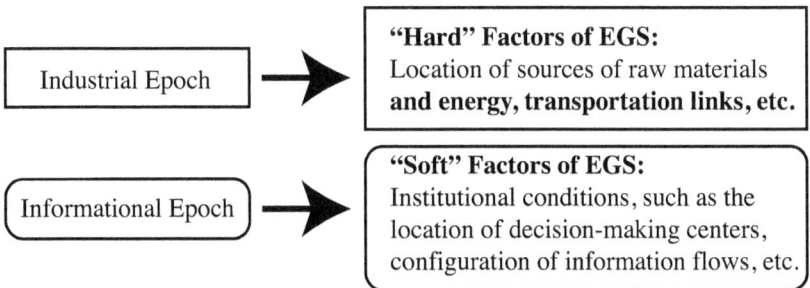

Figure 3.2. | *New Determinants for a City's Economic-Geographical Situation*

roads, exploitation of a new deposit). The EGS institutional factor is more flexible and it can be changed more quickly, but it has extremely important consequences. In the following section, we shall examine the influence of EGS institutional factors on the social-economic development of Gubkinsky and Muravlenko.

For the two monocities we are examining, two factors stand out among the rest: the place of the city in the regional administrative-territorial structure and the role of the city relative to the headquarters of the companies that own the main factory that defines the city's economy. In both cases, the geographic situation of Muravlenko is peripheral, while that of Gubkinsky is quasi-central (that is, it is a local subcenter as defined in Table 3.1).

Before examining the details of the factors that explain the different evolution of the two cities, we provide a brief history of each. Muravlenko appeared in 1984 when Noyabrskneftegaz General Director V. A. Gorodilov created a new settlement, which served de facto as a far-flung part of Noyabrsk. "This construction had not been foreseen by the five-year plan, and therefore had not been financed. But M.K. Mikhailov, head of the Noyabrskneftestroy construction unit, had convinced Gorodilov, Noyabrskneftegaz general director, to begin construction with his own funds and, thanks to those funds, there

Table 3.1. | *Institutional Aspects of Gubkinsky's and Muravlenko's EGS*

Factor	Gubkinsky	Muravlenko
Administrative history	**Earlier administrative independence:** Founded in 1986; in 1988 excluded from the Purpe Rural Council with the formation of the Gubkinsky Rural Council of People's Deputies. Became a city in 1996.	**Long administrative dependence:** Founded in 1984; Muravlenko Rural Council administratively subordinate to the Noyabrsk City Council. Became a city in 1990.
Situation of the main enterprise relative to the main decision-making center	**Subcenter:** Purneftegaz – the most important enterprise of Rosneft, one of the main Russian oil-producing companies	**Periphery subcenter:** Muravlenko oil/gas extraction department subordinate to Noyabrskneftegaz (based in the city of Noyabrsk); Noyabrskneftegaz, in turn, is subordinate to the Sibneft headquarters (Sibneft was bought by Gazpromneft in 2005)
EGS	Local Center	Periphery

was money for construction. In the summer of 1982 builders began working in Muravlenko (Muravlenko City Official Site n.d.)." The main company in the city was Muravlenkovskneft, which was subordinate to Noyabrskneftegaz.

Gubkinsky was founded a little later, in 1986, and farther north. In contemporary conditions, it is separated from Noyabrsk by a three-hour drive (about 240 km); earlier, before the road had been built, this trip took an entire day. The difference in the cities' transport situation was critical in forming the different institutional components of the EGS distinguishing the two cities. The central company in Gubkinsky was Purneftegaz. On 14 July 1986, the Ministry of Oil Industry had created Purneftegaz on the base of Noyabrskneftegaz as a new, independent oil- and gas-producing enterprise that was part of Glavtyumenyeftegaz. After the elimination of the mother company in 1991, Purneftegaz became an independent state enterprise that functioned with this status until July 1993, when, in the course of privatization, it was turned into an open stock company. On 25 September 1995, the company joined Russia's main state oil company Rosneft as a subsidiary.[4]

Thus, by being located close to Noyabrsk, Muravlenko became a kind of "tentacle" development base that was directly connected to the mother city. Businesses in the city for many years had been institutionally dependent on Noyabrsk structures, both production oriented and administrative. It was precisely this peripheral nature in relation to Noyabrsk for many years that shaped the key features of the city's development. For Gubkinsky, which was located 120 km further away, Noyabrsk had much less significance: the difference of 120 km turned into a radical difference for the institutional aspects of EGS.

The administrative history of the cities developed in an analogous way. The Rural Soviet of People's Deputies of Muravlenko for the first six years of its existence was subordinate to the city council of Noyabrsk despite the distance of 120 km between them. The settlement of Gubkinsky, established in 1986, became administratively independent already by 1988.

Over time, the status of the two cities' key companies became less similar within the structure of their parent companies: Purneftegaz gradually lost its independence within the structure of Rosneft, while Muravlenkovskneft became more independent. However, the earlier-laid trajectories of city development, apparently, were stronger than more recent changes in the EGS: the systems of power and the specific social institutions were built according to the initial conditions that had been created in each of the cities.

Thus, the difference in the transport-geographic situation of the cities in the early stages of the development of the territory (and, accordingly, at the stage of building the administrative, organizational, and social relations in the new territory) resulted in the fact that the new cities (and their businesses) took different places in the newly created hierarchy of regional centers—i.e., the differences in the institutional aspects of their EGS were formed. Even though the difference in the transportation-geographical situation has been partially overcome by the development of transportation infrastructure (particularly good quality roads), the earlier established difference in the institutional aspects of the EGS continues to exert influence on the socioeconomic development of both cities. Such "deterministic" models of development are usually called *path dependency*—and, interestingly, in our case path dependency includes both hard and soft factors of the EGS. Accordingly, we conclude, that settlements or cities built farther from earlier centers for developing a certain territory have a greater chance of becoming independent centers than cities developing "in the shadow" of existing centers.

The institutional aspects of EGS, in turn, could consolidate and strengthen the differences in the "hard" factors of the economic-geographical situation; this is precisely what happened in relation to the development of the local railroad network. The remote location of the deposit in the place of the future Gubkinsky settlement required the immediate construction of a separate railroad station—Purpe. Initially the planners suggested building a city, which would serve as the base for exploiting the deposit, in the location of the Purpe township, but due to the complicated geological condition of that place, Gubkinsky's founding was moved 18 km away. Muravlenko until now has not been connected to the railroad: the nearest station, Khanymei, is 80 km from the city and passengers typically use the Noyabrsk center train station.

Gubkinsky Outside Noyabrsk's Social Networks

The institutional aspects of a city's EGS are affected also by factors beyond its physical distance from the company headquarters and other administrative offices. A major role is played by the city's relationship to the most powerful social networks, especially those of the regional and industrial power elites. These networks form strong social ties, like we have seen in Italy, Latin America, and in diaspora networks (such as those supported by the Chinese, Israelis, and Turks, among others).

These networks played an important role in defining the development trajectories for Gubkinsky and Muravlenko: Gubkinsky from the very beginning was outside of the administrative and social networks of Noyabrsk, with whom Muravlenko had maintained close ties. Over the years, there were numerous attempts to gain ownership of Gubkinsky's major oil and gas enterprise, Purneftegaz. From 1998 on there had been an open debate initiated by the Gubkinsky elites to prevent the city's most important property from falling into "the zone of influence" of Noyabrsk. Formally, Purneftegaz had been declared bankrupt and Rosneft's packet of stocks in the company had been sold for $10 million. This action would have allowed the assets to be taken over by the Noyabrsk-based Sibneft, then owned by oligarch Boris Berezovsky. However, a court ordered that the company again be restored to Rosneft. Purneftegaz General Director Aleksei Matevosov was the key figure in the battle seeking to align the company more closely with Noyabrsk elites, particularly those associated with Sibneft, but he was fired from the leadership of the company and subsequently lost influence (Poluektov 1999).

While these property battles remain murky and complex, the ultimate result was that Gubkinsky had weaker ties to Noyabrsk than Muravlenko had, even at the level of the social networks of common people living in these cities. According to research we conducted in 2012, of all students studying in Gubkinsky schools, only fifteen individuals later migrated to Noyabrsk.[5] That figure represents only 0.5 percent of the young people in this age category or only 1 percent of those leaving Gubkinsky. Of all studying in Muravlenko, sixty-eight migrated to Noyabrsk—1.3 percent of all Muravlenko young people or about 3 percent of those leaving Muravlenko.

Authority, Property, and Society: The Gubkinsky "Trio"

Beyond the city's key enterprise, the mechanism by which the socioeconomic development of the city depends on EGS institutional aspects lies in the details of the model of interaction between the municipal authorities and the local society. In making decisions, the authorities in peripheral cities often are oriented on external partners, implementing a "colonial model of authority." The authorities of a more independent "central" city want to let the city function based on its own resources, and, primarily, develop ties with the city's community. This system leads to the formation of a "socially embedded model of authority." In the following discussion we examine the differences between these two models.

The Participation of Society in Adopting Management Decisions

In Muravlenko, under the conditions of a colonial model of management, the system works in an authoritarian manner.[6] Under this system, it is typical for the decision makers to work with their counterparts in Noyabrsk and pay minimal attention to local society. This approach is distinctly visible in the low number of commissions and consultative councils that Muravlenko has created: seventeen as of October 2012, while Gubkinsky has created fifty-four (see Table 3.2).

In addition to these commissions, other institutions of local society are also developing, including trade unions and other societal organizations. In particular, since 1994 trade unions have played a strong role in Gubkinsky. Ruzhitskii explains their importance:

> The privatization of Purneftegaz took place in 1993. By that time, there was already a strong trade union organization at the enterprise, which encompassed the labor collectives of all its subdivisions and expressed the collective opinion of all its workers. In great part thanks to the active position of the united professional committee and the ability of its chairman to defend the interests of northerners at all levels, including with the federal government, a decision had been taken in favor of the workers' collective and the enterprise did not become a part of the no-longer existing firm Sidanko. Instead, the battle for the ownership of Purneftegaz continued for another two years and only in 1995 did it join Rosneft. In 1998, the company went through bankruptcy and the difficult non-payments cri-

Table 3.2. | *Local Community Commissions and Consultative Councils*

Muravlenko	Gubkinsky
Commission on Children's Affairs and Defending their Rights	Commission on Children's Affairs and Defending their Rights
Commission on Observing the Regulations for Employee Behavior and Resolving Conflicts of Interest among Municipal Employees in the City Administration	Commission on Observing the Regulations for Employee Behavior for Employees of Gubkinsky and Resolving Conflicts of Interest
City Society Consultative Council under the aegis of the City Administration	Social Housing Commission

City Interagency Commission on Realizing the YNAO Program to Aid Voluntary Migrants to the Russian Federation of Countrymen Living Abroad

Council on Social and Religious Associations

Commission on Providing Preliminary Expert Evaluations of Consequences of Public Policies

Trilateral Commission on Regulating Social-Labor Relations in the City of Gubkinsky

Commission for Inspecting Individually Built Housing

Commission for Examining the Applications of Citizens Seeking Plots of Land Free of Charge as Property for Individual Housing Construction

Commission for Assisting Poor Families and Individuals to Become Self-Sufficient

Commission for Checking the Correctness of Municipal Fees Paid by Citizens

Coordination Council on Handicapped Affairs

Council on City Construction

City Interagency Council on the Professional Education of Young People

Commission on Ensuring the Participation of the Population of Gubkinsky in the Exercise of Local Government

Coordinating Council on the Development of Small and Medium Sized Business

Coordinating Council on Realizing the Basic Directions of Family and Demographic Policies

Interagency Commission on Organizing Rest, Health, and Work among Youth

Mayor's Council on Young People

Expert Commission on Evaluation Proposals about Determining the Places that Could Harm the Health of Children

Source: Authors' compilation

sis. At that time the trade union organization opposed the presumptuous enterprise leader [Aleksei Matevosov], who had adopted a decision "to improve the financial health of the organization" by reducing salaries by 40 percent, and organized protest actions and successfully defended the workers' interests in court. Moreover, later the "general" [Matevosov] was fired. (Ruzhitskii 2012)

Additionally, the Gubkinsky City Administration regularly holds round tables on a variety of questions and invites a wide range of participants. There are also various consultative and holiday events for local business people and a variety of other events.

Local Culture

Consolidation of the local community and full utilization of its creative potential are connected with the formation of local identity. Elinor Ostrom highlighted the role of consolidation of the local community and clear recognition of its borders as a key factor for the effective management of its common property (the commons) (Ostrom 2010). Former Alaska Governor Wally Hickel had emphasized the role of the "spirit of history" in local society in the economic success of Alaska (Hickel 2004):

> Residents of the state of Alaska should again retrieve the spirit of the pioneers, which they had before the discovery of oil in Prudhoe Bay, when Alaskans thought that they were richer and could imagine themselves in the most unbelievable dream ... This richness was our pioneering spirit. ...
> Following the discovery of oil on the North Slope of Alaska, a wave of new settlers from other states gushed into our state. Some of these new settlers subscribed to a set of values that differed from those of Alaskans. The new arrivals often were attracted by the possibility of finding here high-paying work or carrying out their noble ideas of ecological ethics. Only a few knew the history of the state of Alaska and the majority of these did not understand the fundamental idea of creating a state-owner (Hickel 2004).

In a general sense, identity provides legitimacy for the existing order (Timofeev 2008). Territorial identity, in our opinion, is connected to the legitimization of local community as an independent social and economic actor: recognizing community as a collective subject ("we" as distinct from a collection of separate residents who do not form an integral whole) makes it possible to act in its name, announce the right of the community to resources, comfortable surroundings, participation in decision making, and other acts. In turn, the most important instrument for forming and supporting collective identity is considered to be collective memory,[7] and the most important institution for preserving collective memory is the museum.

The idea of museums as an institution representing collective memory is not new. However, in the last two decades, the representation of the collective memory as a factor in the formation of identity is one of the main issues in Russian museum affairs (Gerasimenko 2012; Svyatoslavskii 2011). Against the background of postindustrial development, the increased role for the creative class, and the destruction of the boundary between work and leisure and between "productive" and "private" spaces, museums are also being transformed: their role as a demonstrator of the collective memory is increasingly moving toward producing exhibits and performances directly on the street.

The museum is even more important in the context of a small city and especially one that is cut off from the main corpus of the country's population. In a small town setting, the museum often becomes the most important institution for integrating the cultural life of the local society and the main space for communication.

The Gubkinsky city leaders intuitively understood the importance of the museum for the formation of the "spirit" of society in the young northern city, even though they apparently were not familiar with contemporary culturological works.[8] Nevertheless the city leaders sought to exploit the greatest potential of the local community. The result was generous financing for the museum: in terms of floor space, the number of exhibits, guides, and researchers, the museum of Gubkinsky is significantly larger than its "counterpart" in Muravlenko. As a result, although the museums of the two cities discussed here (just like the cities themselves) have functioned for almost the same amount of time (the Gubkinsky Museum on the Development of the North opened in 1996, while the Muravlenko Ecological-Regional Studies Museum was founded in 1998), the difference between them is both quantitative and qualitative (see Table 3.3).

The larger number of visitors flowing into the Gubkinsky museum reflects the higher quality of its museum. It is significantly better integrated into the life of the city's community and is an informal communication platform for groups of residents of different ages. The Gubkinsky museum is closely tied to the major enterprises in the city. Its exhibits present detailed information on the industrial activities taking place in the city—in particular, the museum has developed an interactive computer game explaining oil and gas extraction. In return, Gubkinsky's oil and gas companies have provided the resources

Table 3.3. | *Gubkinsky and Muravlenko Museum Indicators*

	Muravlenko Ecological-Regional Studies Museum	Gubkinsky Museum on the Development of the North
Floor space, m²	328.9	733.9 (2006)
Overall number of employees, 2011	8 (including technical personnel)	12 (not including technical personnel)
Number of collection items held, 2011	5,622	8,693
Number of visitors in 2011, thousand individuals	About 12	23.5
Number of visitors in 2011 per 10,000 city residents	3.6	9.1

Sources: Municipal Budgetary Institution of Culture, "Ecological-Regional Studies of the Museum of Muravlenko, Department of Culture, Sport, and Youth Policy of the City of Muravlenko," retrieved 12 April 2013 from http://uksimp.muravlenko.com/ob-upravlenii/podvedomstvennye-uchrezhdeniya/uchrezhdeniya-kultury-i-iskusstva/muk-ekologo-kraevedcheskiy-muzey-g-muravlenko/; Department of Culture, City of Gubkinsky, retrieved 25 December 2012 from http://www.gubadm.ru/gubadm/life/kult/uk/pod/muzey.php; and "Report on Results of Activity for 2011 and Basic Directions for 2012–2015 of the Department of Culture of the City of Gubkinsky in Coordination with the Completion of the Strategic Goals of the Social-Economic Development of the City of Gubkinsky," 2011.

making possible the integration of the museum into the social life of the city through special projects, such as a grant from the firm Gazprom Dobycha Noyabrsk for "Inviting parents to the museum" in 2011. At the same time, the energy company of Muravlenko sponsored only one photo exhibit in 2011. In Muravlenko, the museum's role in leisure time activities of city residents was limited to a small group of members of the "Family Club" (which engaged in needle-craft activities); Gubkinsky's museum was undertaking a range of activities for various categories of visitors, including such projects as "School for future mothers," and "Retirement excitement."

In addition to better support for its museum, Gubkinsky excels Muravlenko in the level of support for cultural activities in general. In this respect, Gubkinsky distinguishes itself not only from Muravleno, but many other larger cities in the YNAO (See Figure 3.3).

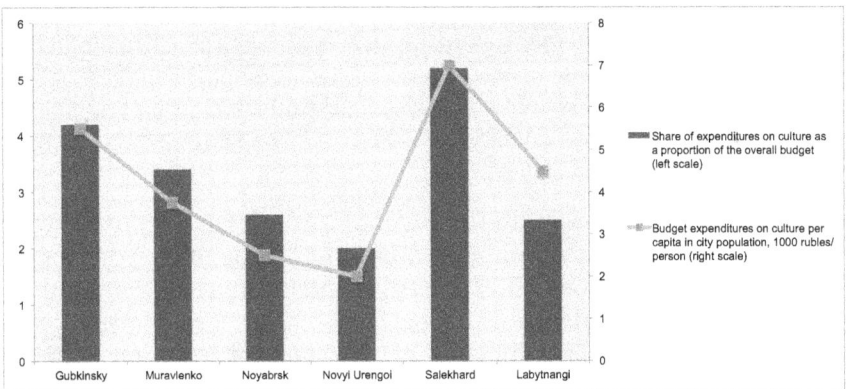

Figure 3.3. | *Expenditures on Culture in Gubkinsky and Other YNAO Cities, 2011*
Source: City Administrations

Local identity and culture are necessary factors in the "embeddedness" of the local community, formation of favorable socioeconomic conditions for living, and "strengthening" the population of the city. In this sense, there are extensive differences between Gubkinsky and Muravlenko.

Gubkinsky authorities define the problem of embeddedness as one of the most important. To support this statement, we provide the following example:

> In contrast to several other more northern cities in the YNAO, where a big secondary housing market is developing, Gubkinsky is creating new buildings and provides housing finance for its residents. Apartments in the Far North are expensive, but the Gubkinsky city administration tries to create the best payment conditions for its citizens in purchasing an apartment. Moving people into new apartments often solves the difficult problem of wooden houses destroyed by fires. The mayor [V. Lebedevich] has even issued an order preventing builders from utilizing wooden construction materials. Lebedevich considers his most important achievement to be that the feeling of temporary life, so common for fly in/fly out settlements [see Chapter 4 in this volume], has begun to dissipate in Gubkinsky. People have stopped feeling cut off from the rest of the world, they have begun to thoroughly settle in this new place, and plan to live here for a long, long time. (Yudina 1999)

In contrast, Muravlenko's mayor considered limiting population growth in the city as his success:

I want to point out that the population of the city of Muravlenko is not growing. Just as we had 37,000 residents earlier, the same situation holds today. In my view, this is one of our serious accomplishments. Northern monoindustry towns should not grow because otherwise their productivity and social structure will deteriorate. We have preserved our structure and that is very good. (Igtisamova 2009)

A policy aimed at limiting population growth, understood exclusively as a function of the number of workers in the oil and gas industry, leads to a reduction in the human potential of the local community. The following quotation from a youth discussion forum, where conversation focused on the mayor who had held this position until the fall of 2010, serves as a striking example of this problem:

I once studied at the Muravlenko branch of Tyumen State University, and during the second year, our esteemed mayor gave lectures. He said something that I could not forget. "Get out of here, there is nothing for you to do here with your education in 'state and municipal management,' you are not needed here. My people are sufficient to work in the city administration. We need workers and oil specialists." And these are the words of a mayor who has an interest in the development and flourishing of the city? (No author No date)

Interest of Young People to Return to Muravlenko and Gubkinsky: Evidence from the Internet

An analysis of Internet forums in Gubkinsky and Muravlenko conducted in January 2012 deepens our understanding of the connections between the colonial or embedded models, the general spirit of the urban community, and the level of business activity in the city.

The Internet is increasingly being used as a source of material for sociological and other social science research (Chugunov 2003; Filippova 2000, 2001). The appearance of the new methods is logical: this most important network for communications provides extensive new material for study. Among the sources the Internet provides are the contents of forums, personal pages put up by users of social media, and the structure of their network of "friends." Numerous social, political, and ecological movements and subcultures use the Internet for communications among themselves, to organize mass actions, demonstrations, and flash-mobs.

Additionally, Internet blogs serve the function of personal diaries. Journal entries have long served, along with in-depth interviews, as material for qualitative analysis in sociology since they make it possible to uncover the motives and mechanisms of many complicated, new, previously unstudied social phenomena; qualitative methods are necessary where scholars want to understand the nature of a phenomenon that was previously unknown, describe new aspects of already known problems in detail, and to uncover the hidden, subjective thoughts and mechanisms of the functioning of social practice that cannot be studied through using mass questionnaires or quantitative data. The material from city forums provides a specific body of written sources, analogous to private diaries, which makes it possible to create a qualitative evaluation of the connections between such phenomena as creativeness, economic expectations, and territorial identity.

A standard thematic coding, conducted in the course of content-analysis (Shteinberg 2009) of comparable groups of statements[9] on the forums of Gubkinsky and Muravlenko shows a varied list of themes discussed in the two cities. Against the background of a large number of common themes, only in Gubkinsky is there a discussion area addressing the theme of returning to the city after graduating from university with the goal of working there; only in Muravlenko are there unprovoked derogatory remarks about the city (see Table 3.4).

Overall, there were eight messages on the theme "Who, if not we?" in the forum "Will you return to Gubkinsky?" Together with the expressions of support for the statements of the other participants, this is 11 percent of the total number of messages on the forum. For the strategy "Get out of this hole!" there were 11 messages (16 percent of the messages in the forum "Will you return to Murka?"). Thus the problem of being passive or holding negative attitudes toward the city are specific to Muravlenko, and this is what distinguishes Muravlenko from Gubkinsky in the area of communications and self-perception in the urban community.

Ultimately the embedded model of power, combined with including the local community in the process of managing the city (through consultative committees affiliated with the local government), and the attention of the authorities to local cultural specifics (such as financing the museum) leads members of society to form an active position ("the feeling of ownership") in relation to their city and a strong territorial identity. By contrast, the colonial model of power and property is connected to widespread feelings in urban society of powerlessness and alienation.

Table 3.4. | *Unique Themes in the Cities' Online Forums*

Topic (informal name)	Description	Examples of messages	Appearances in city forums	
			Muravlenko	Gubkinsky
"Get out of this hole!"	Exclusively negative evaluation of the city (often using profanity), linked to the decision to move away from it	There is nothing to do in this hole... why go there if it only makes you dull	+	−
"Who if not we?"	Active support for the development of the city tied to the well-thought out perspective of returning to the city after completing studies	And then think. If one can return and change life there? You can do what is interesting to you anywhere. It would be hard to prevent this, and our city would be unlikely to. It will not develop on its own, Rosneft will do nothing to help it, but city residents fully could provide good support	−	+
		I have not gone away:) the city is developing, there are new opportunities:) it is interesting to develop with one's home town:) the city is not apartments and various buildings, but WE! PEOPLE!:)		

Source: Vkontakte.ru (see endnote 9 for details)

The Role of Small Business

Returning to the topic of small business, with which this chapter began, we can now make the argument that the strong feeling of local identity and local patriotism has provided Gubkinsky with "fertile soil" for developing the city's endogenous economic potential. In a small city, the local economic potential primarily consists of developing small and medium business. Thus the most important critical indicator of the local authorities' orientation to relations with the local community or with external agents (Noyabrsk) is its relationship with small business. The embedded model of Gubkinsky, focused on the maximum development of local resources, highlights its effective support for small business. In contrast, in Muravlenko, small business develops mainly despite the existing system of relations between the authorities and local community.

If local identity provides legitimacy for the development of local society, then small business is its direct mechanism. As Figure 3.4 shows, leaders and outsiders among the cities of YNAO in terms of their level of support for small business, on one side, and the number of museum employees per capita, on the other side, practically coincide.

Both the supply of museum employees and support for small business have practically "collapsed" in Muravlenko and Novyi Urengoy, are slightly better in Noyabrsk, but are at the peak in the capital Salekhard, Labytnangi (which is close to the capital), and Gubkinsky.

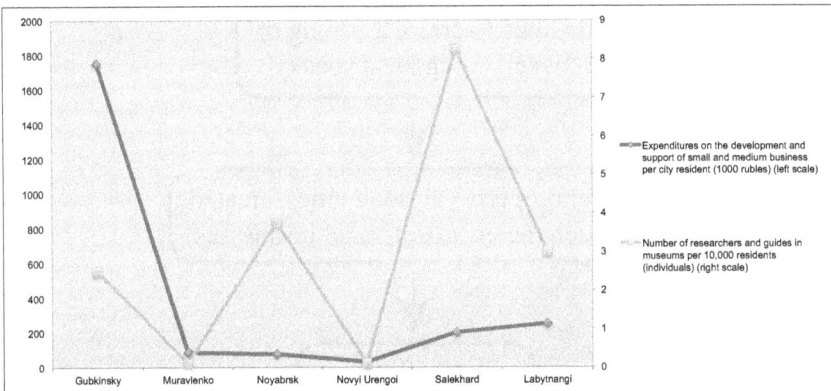

Figure 3.4. | *Small Business and Museum Development in YNAO Cities*
Source: City Administrations

In the conditions of the colonial model, the entire structure of the local political authorities in Muravlenko reduced the window of opportunity to diversify the city economy. In contrast, in the conditions of the embedded model in Gubkinsky, the local authorities actively encouraged the development of the productive forces of small business, which became an important instrument of diversification for this monocity (see Figure 3.5).

Conclusion

In both cities, the model of power, local cultural milieu, entrepreneurial energy, activity of social organizations, and local identity all had influential impacts on each other. At the same time, the trigger for the

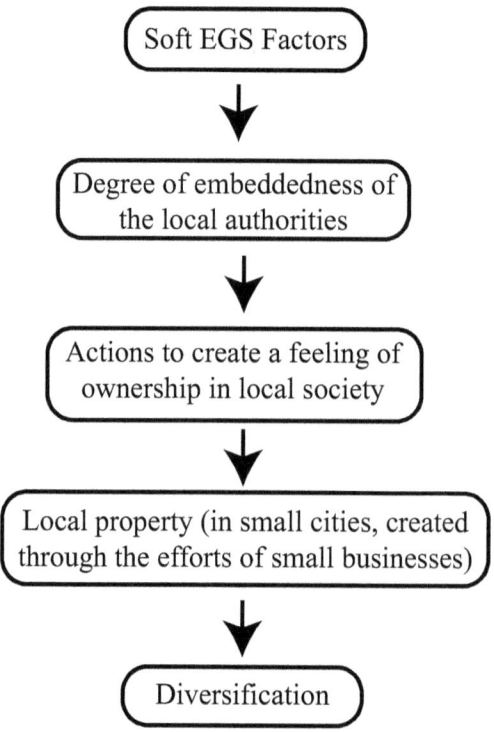

Figure 3.5. | *Impact of the Economic-Geographical Situation (EGS)*

formation of their specific model of interaction (whether embedded or colonial) has been the institutional aspects of the economic and geopolitical situation of the cities (respectively, central or peripheral). Consequently, the possibilities for diversification of the economy were determined by the specific local interactions in the system of "authority-property-society."

Nadezhda Yu. Zamyatina is Senior Research Fellow in the Department of the Social Economic Geographies of Foreign Countries, Geography Department, Moscow State University.

Aleksandr N. Pelyasov is the Director of the Center of the North and Arctic Economy, a state research institution of the Council for the Study of Productive Forces, Ministry of Economic Development, Russian Federation, and the Russian Academy of Sciences.

Notes

1. The social organization of Gubkinsky Union of Entrepreneurs, retrieved 12 April 2013 from http://www.gubadm.ru/gubadm/economy/msb/gsp.php.
2. "The Union of Entrepreneurs Has Been Created in the City of Muravlenko," *RIA-Novosti Ural* 15 November 2002. Retrieved 12 April 2013 from http://ural.ria.ru/economy/20021115/39611.html.
3. Possibly there are additional explanatory factors. Small business often strongly depends on local consumer demand. Gubkinsky has a higher average salary than Muravlenko—maybe small businesses there have been stimulated by the local demand? Additionally, not everything is so simple with the extraction of hydrocarbons: Gubkinsky began to produce gas with the company Purgaz in addition to oil in 1998, earlier than in Muravlenko. Gas production in this area is significantly more profitable than oil. Muravlenko for many years produced only oil, and gas production only began in 2011. One could hypothesize that Gubkinsky's larger consumer market was driven by gas workers. But this was not the case. First, Gubkinsky's gas sector employs only 200 personnel, though they are indeed well paid. Second, the weighted indicators of consumer buying power for the development of small business in Gubkinsky significantly exceed the same indicators in Muravlenko. Third, and most important, the small business of Gubkinsky includes not just trade and services, which depend heavily on the size of the consumer market, but also business services for enterprises that make paving slabs and other construction materials and servicing trucks driving along the same road on which Muravlenko is located. Thus, the difference in salary levels between Gubkinsky and Muravlenko is in no way the main differentiating factor for their development.

4. OOO RN-Purneftegaz, retrieved 21 July 2014 from http://progubkinskiy
.ru/%D1%81%D0%BF%D1%80%D0%B0%D0%B2%D0%BE%
D1%87%D0%BD%D0%B8%D0%BA/457-%D0%9E%D0%9E
%D0%9E-%D0%A0%D0%9D-%D0%9F%D1%83%D1%80%D0%
BD%D0%B5%D1%84%D1%82%D0%B5%D0%B3%D0%B0%D
0%B7.
5. To conduct this research, we examined 3,263 personal pages in VKontakte belonging to young people aged 20–29 years old who at one point studied in Gubkinsky and 5,221 pages of those who at one point studied in Muravlenko.
6. V.A. Bykovsky was the mayor of Muravlenko from 1997 to 2010. Before becoming the mayor, he was a member of the YNAO State Duma, and earlier, a member of the Tyumen Oblast and Yamal-Nenets Okrug Council of People's Deputies. Before entering politics, he worked in the oil and gas extraction sector in Samara Oblast, Udmurtiya, Surgut, Nizhnevrtovsk, and Raduzhnyi. Working in the government offices at the *oblast* and *okrug* levels helped him to form a "colonial" style of management in Muravlenko. The mayor of the settlement and later city Gubkinsky, V.V. Lebedevich has held this job since 1993. Earlier as a commissar for the Belarussian Komsomol brigade named Molodogvardeets, he participated in the construction of the railroad from Purpe. In 1981–1988, he worked in the Purpe Raion Committee of the Communist Party of the Soviet Union. Beginning in 1988 he was the deputy chairman and then chairman of the executive committee of the Purovsky Raion Council of People's Deputies. In 1992, he became deputy mayor of Gubkinsky. Thus, practically the entire career path of the future mayor of Gubkinsky has been connected with the Purovsky Raion and literally the area around Gubkinsky, while the career path of the future mayor of Muravlenko has been associated with a larger territory, including big cities. Their staying in power in both cities for so long makes it possible for us to consider the model of authority in both cities not changing over a long period of time.
7. Other terms used include "social memory" and "cultural memory." See: Middleton and Edwards 1999 and Misztal 2003. The German scholar Jan Assmann (1992) uses the term "cultural memory," which is not popular in English language "memory" studies.
8. Irkutsk Sociologist M. Rozhansky pointed out to us that in young Siberian cities, innovative educational programs in schools appear, in particular, around the topic of regional studies (*kraevedeniya*) since a functional "cultural memory" is relevant for their pioneering societies.
9. For comparative analysis, we chose thematically close discussions in the youth forums on the VKontakte site: the theme "Who will return to Gubkinsky?" in the group "Gubkinsky—dream city" (http://vk.com/feed#/topic-29017_217511; 3,531 participants in the group; 67 messages on the topic); the theme "Will you return to Murka?" in the group "We're going to Murka!!!" (http://vkontakte.ru/feed#/topic-614041_646050, 1,844 participants and 69 messages; the theme "One word about Gubkinsky" in the group "Gubkinsky—Dream City" (27 messages); theme "Do you love Gubkinsky?" in the group "Gubkinsky People" (http://vk.com/feed#/

topic-25981350_24304245; 330 participants; 27 messages); topic "Muravlenko—city or village" "Everyone from Muravlenko come here!!!" (http://vkontakte.ru/topic-614041_646050#/topic-4510135_7642410; 544 participants; 58 messages; topic "What are your impressions of Muravlenko?" in the groups "Muravlenko (YNAO) (http://vkontakte.ru/topic-147384_152237; 1,436 participants, 172 messages).

References

Animitsa, E.G. and N.V. Novikova. 2009. "Problemy i perspektivy razvitiya monogorodov Rossii [Problems and prospects for the development of Russia's monocities]," *Upravlenets* 1–2: 46–54.
Assmann, Jan. 1992. *Das kulturelle Gedaechtnis. Schrift, Erinnerung und politische Identitaet in fruehen Hochkulturen* [Cultural memory. Scripture, memory and political identity in early civilizations]. Munich: C.H. Beck.
Chugunov, A.V. 2003. Sotsiologiya Interneta: Sotsial'no-politicheskie orientatsii rossiiskoi internet-auditorii [Sociology of the Internet: Sociopolitical orientations of the Russian Internet audience], 2nd ed. St. Petersburg: Filologicheskii fakultet [St. Petersburg State University].
Filippova, T.V. 2000. "Interaktivnaya kommunikatsiya v empiricheskoi sotsiologii [Interactive communication in empirical sociology]," in *Sociology*. Moscow: Russian State Humanities University.
———. 2001. "Internet kak instrument sotsiologicheskogo issledovaniya [The Internet as an instrument of sociolgocial research]," *Sotsiologicheskie issledovaniya* 9: 115–22.
Gerasimenko, E.E. 2012. "Muzei v institutsionalizatsii sotsial'noi pamyati [The museum in the institutionalization of social memory]," in *Culturology*. St. Petersburg, Russia: St. Petersburg State University.
Hickel, Wally. 2004. *Problemy obshchestvennoi sobstvennosti. Model Alyaski—Vozmozhnosti dlya Rossii?* [Problems of social property. The model of Alaska—Possibilities for Russia?] Moscow: Progress.
Igtisamova, A. 2009. "S lyubov'yu k gorodu (Interv'yu c V. A. Bykovskim) [With love for the city (Interview with V.A. Bykovsky]." *Nash gorod*.
Kuznetsova, G. Yu. and V. Ya. Lyubovnii. 2004. *Puti aktivatsii sotsial'no-ekonomicheskogo razvitiya monoprofil'nykh gorodov Rossii* [Methods for activating the socioeconomic development of single-profile cities in Russia]. Moscow: Moskovskii obshchestvennyi nauchniy fond.
Middleton, D. and D. Edwards eds. 1999. *Collective Remembering*. London: Sage.
Misztal, Barbara A. 2003. *Theories of Social Remembering*. Philadelphia, PA: Open University Press.
Muravlenko City Official Site. No date. "Istoriya goroda [History of the city]."
No author. No date. "Ya kogda uchilas' ... [I once studied ...]." Muravleno City Official Site. Internet Forum. Message 2450.
Ostrom, Elinor. 2010. *Governing the Commons: The Evolution of Institutions for Collective Action*. Cambridge: Cambridge University Press.
Poluektov, Nikolai. 1999. "Kreditory 'Purneftegaza' poshli po mirovuyu [Creditors of Purneftegaz agreed to a peaceful resolution]." *Kommersant*.

Ruzhitskii, I. 2012. "Rol' klyuchevoi figury [The role of the key figures]." *Gubkinskaya nedelya*.

Shteinberg, I. ed. 2009. *Kachestvennye metody. Polevye sotsiologicheskie issledovaniya* [Qualitative methods. Sociological field research]. St. Petersburg: Aleteiya.

Svyatoslavskii, A.V. 2011. "Sreda obitaniya kak sreda pamyati: k istorii otechestvennoi memorial'noi kul'tury [Means of living as a means of memory: Toward a history of the domestic memorial culture." In *Culturology*. Moscow: Moscow Pedagogical State University.

Timofeev, I.N. 2008. *Politicheskaya identichnost' Rossii v postsovetskii period: al'ternativy i tendentsiya* [The political identity of Russia in the post-Soviet period: alternatives and tendencies]. Moscow: MGIMO.

Yudina, Lyudmila. 1999. "Vdali ot Bol'shoi Zemli [Far from the mainland]." *Trud*.

Zubarevich, Nataliya. 2012. "Sotsial'naya differentsiatsiya regionov i gorodov [Social differentiation of regions and cities]," *Pro et Contra* 16(4–5): 135–52.

Section II

Migration Trends in Russian Arctic Cities

CHAPTER FOUR

Boom and Bust
Population Change in Russia's Arctic Cities

Timothy Heleniak

Introduction

This chapter provides context for examining sustainability in Russia's Arctic cities by examining patterns of population change over the past two decades.[1] These demographic statistics are an important starting point for the examination of changing spatial patterns of economic activity, changes in settlement patterns and social networks, and the construction or contraction of urban settlements in Russia's Arctic.

The manner in which the Soviet Union's centrally planned economy developed the resources of its Arctic and northern regions differs considerably from the development path that Russia is pursuing. Some of the greatest human-environment battles that the Soviet Union undertook during its seven decades of existence were in the development of the Arctic. While not completely answering the question of whether the current development practices in the Russian Arctic are sustainable, this chapter at least partially addresses the issues by examining the vital subject of population change. Whether the state, regions, cities, companies, or individuals are determining population change, it is a crucial indicator of sustainability.

The downsizing of the population of the Russian Arctic in the post-Soviet period has not been one of universal decline, but rather significant shrinkage in most regions and settlements combined with growth in others. The reason is that most of the natural resources critical to Russia's economic growth are located in these regions. The broad question addressed in this chapter is whether the current size of the population of the Russian Arctic is sustainable. From a national standpoint, this means whether there are too many or too few people in large urban settlements in the Russian Arctic. Within each Arctic city, can the current infrastructure support the current population size, or is there excess infrastructure with population decline and what should be done about it? For companies operating in the Russian Arctic, is the size of the labor force optimal to maximize profits? How will rapidly changing climate across the Russian Arctic impact the population residing there, and how will these people impact the climate?

The chapter starts by offering a definition of the Russian Arctic and then proceeds to broadly discuss population change there before moving to finer levels of geographic disaggregation with a focus on urban areas in the Arctic. Subsequent sections examine urban-rural population change and population change in the regional centers and largest northern settlements. The conclusion examines the importance of population changes for urban sustainability in the Russian Arctic.

Defining the Russian Arctic

Like the term *sustainability*, the terms *Arctic* and *North* are often used but not always precisely defined. The Arctic or the North can be defined based on latitude, remoteness, climate, permafrost extent, population density, or other factors. For the purpose of this chapter, which uses statistical data on populations, both the Arctic and North within Russia need to be precisely defined because data are presented for specific geographic regions.

For planning, economic development, statistical, and other purposes, the Russian government defines two different types of northern regions—the Far North (*Krainyy Sever*) and regions equivalent to the Far North (*mestnosti priravnennyye k rayonam Krainego Severa*). The entire territory of ten regions are classified as being in the Far North—Nenets Autonomous Okrug, Murmansk Oblast, Yamal-Nenets Autonomous Okrug, Taimyr Autonomous Okrug, Evenki Autonomous Okrug, Republic of Sakha (Yakutia), Chukotka Autonomous Okrug, Kamchatka Oblast, Koriak Autonomous Okrug, and Magadan Oblast. The Russian government assigns fifteen regions to the Far North on the basis that all or a majority of their territory is classified being in the Far North. In addition to the ten regions listed above, the following are also classified as the Far North—Republic of Karelia, Komi Republic, Arkhangel'sk Oblast, Khanty-Mansiy Autonomous Okrug, and Sakhalin Oblast. The city of Norilsk is also included in this definition of the North.[2] In 1989, these regions encompassed 54 percent of the territory of Russia but only 6.4 percent of the country's population.

Russia also defines certain regions as being Arctic, which is a subset of those defined as North. According to this definition, the Russian Arctic includes the territory of Murmansk Region; the Nenets, Chukchi, and Yamalo-Nenets Autonomous Regions; the municipal formation of Vorkuta (Komi Republic); the municipal district of Norilsk; as well as several areas of Yakutia; two districts of Krasnoyarsk Territory; and municipalities of Arkhangel'sk Region (Marinelink.com 2014). This is similar to the definition of the Arctic in Russia used in the *Arctic Human Development Report* (Arctic Council 2004: 17–18). In that report, the Arctic in Russia includes the Murmansk Oblast; the Nenets, Yamal-Nenets, and Chukotka *okrugs*; Vorkuta city (Komi Republic); the cities of Norilsk and Igarka (Krasnoyarsk Kray); the Taymyr (Dolgan-Nenets) Okrug; and thirteen districts in northern Yakutia. The only difference between these definitions is the inclusion of municipalities in the Archangelsk Oblast.

There are several issues that make analyzing population change in either the Arctic or North problematic. The first is that the definition cuts across *oblast*-level boundaries, yet most Russian population data are only presented at the *oblast* level. This difficulty can be overcome by including all of a region into the category of the North if most of its territory lies in the North. Thus, in this chapter, the North includes the fifteen regions defined above. A related issue is that data on the cities or *rayon* level (third-order administrative units after federal districts and *oblasts*) are scarce and sometimes only data on total population are available and not data on population change or characteristics of the population. A third issue is that three of the northern autonomous *okrugs* were recently abolished and subsumed into their parent regions as part of a larger elimination of autonomous *okrugs*. This change occurred to Taimyr and Evenki autonomous *okrugs*, which became districts within the Krasnoyarsk Kray, and the Koryak autonomous *okrug*, which became a district within the renamed Kamchatka Kray. These combinations of regions are not a major issue for analysis of regional and city populations in the Russian North or Arctic since these are all quite sparsely populated areas lacking any major cities. When analyzing total and urban population change, a distinction will be made between the Arctic and the North.

Migration and Population Change in the Russian North

Our analysis of changing settlement patterns across the Russian North will start with a broad overview and then increase the level of geographic granularity. The level of spatial resolution makes a difference as there is not one northern economy, but many, as the Russian North is simultaneously both under- and overdeveloped. This analysis is based on data from the 1989, 2002, and 2010 population censuses conducted in Russia as well as annual data on births, deaths, and migration.[3] In some cases, data from previous censuses are used, including going back to the last Tsarist census in 1897. Like the population of Russia, the population of the Russian North is unevenly distributed. Of the 7.6 million persons living in the Russian North, two-thirds resided in four regions of the European North, the Karelian and Komi Republics, the Arkhangel'sk and Murmansk *oblasts*, and the Khanty-Mansiy Okrug of West Siberia. Most of the North and Arctic east of West Siberia are quite sparsely populated, with Yakutia and the city of Yakutsk being the only sizeable region and settlement. Most of the autonomous *okrugs* have quite small populations of less than 50,000.

The population of the entire Russian North declined by 20 percent between 1989 and 2013, from 9.4 million to 7.6 million (Table 4.1). Migration has been the main driving force of population change over the period of the economic transition following the collapse of the Soviet Union, with a 22 percent population decline from migration. There was a slight natural increase (the difference between births and deaths) in the population of 2.7 percent at a time when the population of the country was experiencing significant natural decrease due to a steep decline in fertility and rise in mortality to a population that was old and aging rapidly. The reason that the population of the North continued to experience natural population increase was due to its younger age structure, with relatively more people in the young working age groups that are also the higher fertility ages.

Pointing to the fact that two distinctly northern economies have developed during the post-Soviet era, only two northern regions, the Khanty-Mansiy and Yamal-Nenets *okrugs*, have had population increases since 1989. These are the two oil and gas producing regions of Russia. For the Khanty-Mansiy region, its population growth of 24 percent consisted of a natural increase of 19 percent and positive net migration of 4 percent. For Yamal-Nenets, its growth of 10 percent consisted of natural increase of 21 percent offset by outmigration of 12 percent. These two regions had by far the highest percent natural increase in the North because of the young age structure of

Table 4.1. | *Population Trends in the Russian North, 1989–2013 (beginning-of-year; in thousands)*

Region	Total population 1989	Total population 2013	Percent change, 1989–2013 Total	Percent change, 1989–2013 Natural increase	Percent change, 1989–2013 Migration	Absolute change, 1989–2013 Total	Absolute change, 1989–2013 Natural increase	Absolute change, 1989–2013 Migration	Intercensus percent change 1989 to 2002	Intercensus percent change 2002 to 2010
RUSSIAN FEDERATION	147,022	143,347	-2.5	-8.4	5.9	-3,675	-12,307	8,632	-0.9	-1.9
The North	9,466	7,598	-19.7	2.7	-22.4	-1,868	255	-2,123	-14.6	-5.3
Karelian Republic	790	637	-19.4	-11.0	-8.4	-153	-87	-66	-8.7	-10.9
Komi Republic	1,251	880	-29.7	-1.5	-28.2	-371	-19	-352	-17.7	-12.7
Arkhangel'sk Oblast	1,570	1,202	-23.4	-7.5	-15.9	-368	-118	-250	-14.0	-9.3
Nenets Autonomous Okrug *	54	43	-20.4	7.7	-28.0	-11	4	-15	-24.3	2.7
Murmansk Oblast	1,165	780	-33.0	-1.7	-31.3	-385	-20	-365	-22.3	-12.3
Khanty-Mansiy Aut. Okrug *	1,282	1,584	23.6	19.1	4.4	302	245	57	10.1	8.9
Yamal-Nenets Aut. Okrug *	495	542	9.5	21.1	-11.7	47	105	-58	1.6	4.3
Taymyr Autonomous Okrug *	56	34	-38.5	8.3	-46.8	-22	5	-26	-30.1	-12.1
Evenki Autonomous Okrug *	25	16	-35.0	7.8	-42.7	-9	2	-11	-29.0	-8.4
Sakha Republic (Yakutia)	1,094	956	-12.6	14.5	-27.1	-138	159	-297	-13.0	0.7
Chukotka Autonomous Okrug	164	51	-68.9	5.1	-74.0	-113	8	-121	-66.3	-7.8
Kamchatka Oblast	472	320	-32.2	1.0	-33.2	-152	5	-157	-23.4	-11.0
Koryak Autonomous Okrug *	40	19	-53.1	-0.8	-52.3	-21	0	-21	-35.8	-27.0
Magadan Oblast	392	152	-61.2	0.7	-61.9	-240	3	-243	-52.2	-16.7
Sakhalin Oblast	710	494	-30.4	-4.5	-25.9	-216	-32	-184	-22.2	-10.0

Sources: Rosstat, annual demographic yearbooks and census results from 1989, 2002, and 2010.
Note: Data for the Taymyr, Evenki, and Koryak okrugs are for 1989 to 2011.

their populations, which result from having such large populations of migrants.

The breakup of the Soviet Union, liberalization of the society including freedom of movement, and the shift from a centrally planned to a market economy caused a shift in the direction of migration in the North, from moderate in-migration in the 1980s to rather large scale out-migration in the post-Soviet period. By 1990, all northern regions had more people leaving than arriving and this trend has continued past 2000, albeit at much lower rates than the early 1990s. The year of the greatest out-migration was 1992, the first year of the economic reforms and the year that prices were liberalized when the market cost of living in the northern periphery began to be felt. Over the entire period, all northern regions except for the Khanty-Mansiy Autonomous Okrug have experienced out-migration. Eleven of the fifteen northern regions have had one-quarter or more of their populations migrate out since 1989. The only exceptions in addition to the Khanty-Mansiy Okrug were the Yamal-Nenets Okrug and the Karelian Republic and Arkhangel'sk *oblasts*, the latter two being closer and better connected to central Russia. The rates of out-migration increased to the east and in regions with smaller populations. At the extreme are Magadan, which saw an out-migration of 62 percent of its population, and Chukotka, from which nearly three of every four persons migrated out causing the population to fall from 164,000 in 1989 to just 51,000 currently.

Population decline was less in the 2000s than in the 1990s; nearly all northern regions stabilized and a number of northern regions actually had population increases. There was a period in the late 1990s and early 2000s when there was natural decrease in the North, but in the mid-2000s, this trend shifted back to natural increase. Overall, between the 1989 and 2002 censuses, the population declined by 14.6 percent and between the 2002 and 2010 censuses the decline continued, but had slowed to 5.3 percent. For example, Chukotka's population, which had declined by 66 percent during the first intercensus period from 164,000 to 55,000, only declined by another 5 percent to 50,000 in the 2010 census and seems to have stabilized at about this level. The 2000s in Chukotka was the period when oligarch Roman Abramovich was governor and investing large amounts in the region, including some to resettlement programs which had few takers. Likewise, the population of Magadan, which declined by more than half from 392,000 to 187,000 in the first intercensus period, only declined by 17 percent to 156,000 in 2010. The region's population continues to decline, but slowly.

In 2009, the population of Russia experienced growth for the first time since the early 1990s and in 2012, the number of births was only slightly less than the number of deaths. If present trends continue, Russia is expected to have its first natural increase in the post-Soviet period. In 2012, the population of the North declined by only 8,000 persons or 0.1 percent, the smallest rate of population decline in the post-Soviet period. The populations of the Nenets, Khanty-Mansiy, and Yamal-Nenets *okrug*s were all growing moderately. The populations of the Evenki Okrug, Sakha Republic, Chukotka Okrug, Kamchatka Oblast, and Sakhalin Oblast appear to have stabilized or are experiencing no change. The populations of the other northern regions continue to decline albeit at rather low rates.

While the direction of net migration for the North and most northern regions has been negative for most of the past two decades, the direction of the migration flows has hardly been unidirectional, as there have been large flows both from and to the North. The northern regions have always had higher rates of migration turnover than the rest of Russia, meaning that more people are moving to, from, and within the North than the rest of the country. In 1993, near the peak year of net out-migration, there was a net out-migration from the North of 112,000 persons, which consisted of an in-migration of 155,000 and an out-migration of 267,000 (Rosstat, selected years).[4] Put differently, for every 10 persons leaving the North, 6 people migrated to the region. By 2009, the ratio of in-migrants to out-migrants had risen to 8 to 10. These flows consisted of an in-migration of 133,000 and an out-migration of 162,000 for a net out-migration of 29,000. Thus, while there were more people migrating away from the North than to the region, for many the North is still an attraction, at least temporarily. As was the case during the Soviet period, the population of the North is rather footloose and does not have a strong attachment to place, which has implications for social sustainability (Heleniak 2009).

Urban-Rural Population Change in the Russian North

The Soviet Union was a land of large cities and the Russian North even more so than the rest of the country. Forty percent of the population of the North resides in the sixteen northern cities with a population of 100,000 or more. The urban population consists of those living in cities, towns, and urban-type settlements. These are settlements with 12,000 or more inhabitants or where not less than 85 percent are workers, employees, and family members, though this criterion differs among regions. Hill and Gaddy observed that Russia's population is concentrated in cities with few physical connections between them (Hill and Gaddy 2003). This applies even more so to distant northern cities. The Soviet Union underwent one of the most rapid urbanizations in world history as a result of the industrialization of the 1920s and 1930s. This urbanization took place even faster in the east and north than in the rest of the country.

Because the economic structure of the north is based primarily on resource extraction and given the small size of the agricultural sector due to climatic conditions, the north was more urbanized than the rest of Russia, with 79 percent of the northern population classified as urban in 1989 against 74 percent for the entire county. There were differences among northern regions in the share of their populations classified as urban based primarily on the size of the resource-extraction sector. In some of the smaller northern homelands of Siberian natives, which lacked industrial raw materials, large urban settlements were never constructed and their populations remained primarily rural. At the other extreme, in northern regions such as Murmansk, the home of the North Sea Fleet and a number of resource extraction settlements, and the Khanty-Mansiy Autonomous Okrug, the center of the Russian oil sector, over 90 percent of the populations were classified as urban.

For Russia, the urban population declined by 3.1 million (3 percent) between 1989 and 2010, primarily because the large excess of deaths over births was concentrated in urban areas. In the north, urban areas declined by 1.3 million (18 percent) and rural areas slightly more so by 0.4 million (22 percent). The urban population decline in the north can be attributed to the same trend of more deaths than births but also to the fact that out-migration took place primarily from urban areas. The absolute size of the urban population declined in all but two of the fifteen northern regions, the oil and gas regions of west Siberia where there has been considerable investment and economic growth (Figure 4.1). The decline in the urban population in the North of 1.3 million consisted of an increase of 312,000 in Khanty-Mansiy and Yamal-Nenets and a decline of 1,639,000 in the other northern regions. Thus, all urban growth took place in these two regions while there was considerable contraction of the urban populations and presumably urban space in the rest of the North. In terms of urban sustainability, there are two issues. The first is if the environment, much of it consisting of continuous or discontinuous permafrost, can accommodate the growing urban infrastructure in these two regions. The second is what needs to be done with remaining housing and urban infrastructure in the settlements that have had significant contraction of their populations, or complete abandonment.

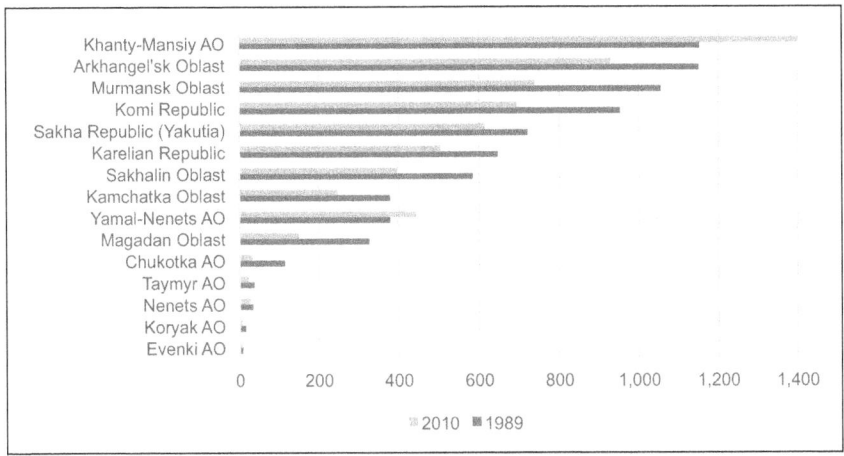

Figure 4.1. | *Size of the Urban Population in the Russian North, 1989 and 2010 (in order by size of urban population in 1989)*
Source: Rosstat, 1989, 2002, and 2010 census results.

The percent urban in a region can change because of three factors: natural increase or decrease, migration, or the reclassification of areas from urban to rural or vice-versa. Overall, the percent urban in the North increased slightly from 80 to 81 percent between 1989 and 2010 while the percent urban in Russia stayed the same at 74 percent. Among northern regions, the general pattern was increases in rates of urbanization in the European North and West Siberia, with the Yamal-Nenets Okrug having the highest rate of increase from 78 to 85 percent of the population. Most northern regions to the east of West Siberia had decreases in percent urban with one notable exception. In Magadan, the urban population declined by slightly more than half from 326,000 in 1989 to 150,000 in 2010, while the rural population declined spectacularly from 60,000 to just 7,000 over the same period. Two-thirds of the total and urban population resides in the regional center of Magadan, with much of the rest of the large region very sparsely populated.

Population Change in the Regional Centers in the Russian Arctic

In Russia, the creation and development of regions and regional administrative centers took place differently than in other countries, such as the United States, where the states were created and then there were battles for which city would become the state capital (Montes 2014). The US pattern often means that the largest city in terms of population and economic size is not the political capital of the state. In Russia, when *oblasts* and ethnic homelands, such as autonomous *oblasts* and *okrugs*, were created in the early Soviet period to replace the tsarist *gubernayas*, they were usually drawn around the largest population settlement in the region, which then also became the political and administrative center. In the North, for all except Khanty-Mansiy and Yamal-Nenets, the administrative center is the largest city and usually the center of economic activity (Table 4.2). In the Khanty-Mansiy Okrug, there are two cities with populations larger than the 80,151 people in the *oblast* center of Khanty-Mansiysk: Nefteyugansk (122,855) and Nizhnevartovsk (251,694). In Yamal-Nenets, there are also two cities larger than the administrative center of Salekhard (42,845): Novyy Urengoy (104,107) and Noyabrsk (110,620). In these two regions, oil and gas developed in and around cities other than the administrative centers.

The spatial distribution of the Russian economy and population was memorably described as "Archipelago Russia" (Dienes 2002). That article pointed out how the geography of Russia was divided into an integrated metropolitan economy centered around Moscow and a vast depressed hinterland, the North and the Arctic. The sheer size of Russia, lack of infrastructure, and environmental obstacles prevent the full incorporation of this periphery into the socioeconomic web of the central portion of the country. Three-quarters of the land mass can be considered a classic hinterland with only a few of the oil-, gas-, and metal-producing centers well integrated into the national and global economy.

The impact of the economic transition was felt in this northern and Arctic periphery at two scales. The first was national as witnessed by the downsizing of the population of the periphery as a result of both voluntary and state-induced closure of settlements and programs designed to assist with migration from the North towards central Russia. The second was contraction of economic activity and population into the regional centers and a few other large urban settlements. To further the analogy of the archipelago, during the economic transition, the waters rose and flooded much of the North, causing the population to concentrate into the few remaining sizeable urban and regional centers. The reason for this is that the cost of economic activity in smaller settlements became prohibitively expensive with the withdrawal of many subsidies to the North, especially on transport. The cost of living rose considerably and public services declined. The "bright lights" of the regional centers, while nowhere near as bright as Moscow, were

Table 4.2. | Population of the Oblast Centers in the Russian North, 1989–2010 (thousands)

Region	Oblast Center	1989	2002	2010	Change 1989-2010 ths.	Change 1989-2010 percent	Population change of region, 1989-2010	Oblast center as share of total oblast population 1989	Oblast center as share of total oblast population 2002	Oblast center as share of total oblast population 2010	Percentage increase or decrease
Karelian Republic	Petrozavodsk	270	267	262	-8	-3	-18	34	37	41	7
Komi Republic	Syktyvkar	219	230	235	16	7	-28	18	22	26	9
Arkhangel'sk Oblast	Arkhangel'sk	417	356	349	-68	-16	-22	27	26	28	2
Nenets Autonomous Okrug	Nar'yan-Mar	20	19	22	1	7	-22	37	46	52	14
Murmansk Oblast	Murmansk	472	337	307	-165	-35	-32	41	37	39	-2
Khanty-Mansiy Aut. Okrug	Khanty-Mansiysk	35	54	80	45	126	20	3	4	5	2
Yamal-Nenets Aut. Okrug	Salekhard	32	37	43	11	33	6	7	7	8	2
Taymyr Autonomous Okrug	Dudinka	37	27	22	-14	-39	-39	65	70	64	-1
Evenki Autonomous Okrug	Tura	7	6	6	-2	-25	-35	30	33	34	4
Sakha Republic (Yakutia)	Yakutsk	188	210	270	82	43	-12	17	22	28	11
Chukotka Autonomous Okrug	Anadyr'	18	9	13	-5	-28	-69	11	16	26	15
Kamchatka Oblast	Petropavlovsk-Kamchatskiy	273	198	180	-93	-34	-32	58	55	56	-2
Koryak Autonomous Okrug	Palana	4	4	3	-1	-27	-53	11	15	17	6
Magadan Oblast	Magadan	152	99	96	-56	-37	-60	39	53	62	23
Sakhalin Oblast	Yuzhno-Sakhalinsk	156	175	182	26	16	-30	22	32	37	15

Source: Rosstat, 1989, 2002, and 2010 census results.
Sources and notes: Population totals are for permanent population (*postoyannoye naseleniye*). Figures for city populations may include an element of boundary change. 1989: Unless otherwise noted, Goskomstat Rossii, Goroda Rossiyskoy Federatsii, Moscow, 1995, pp. 12–13. For smaller regions and cities (noted in grey), population totals are from 1990 and are from Gokomstat RSFSR, Chislennost' naseleniya RSFSR po gorodam, rabochim poselkam i rayonam na 1 yanvarya 1990 g., Moscow, 1990. 2002 population totals (October 2002 population census): Goskomstat Rossii, O predvaritel'nykh itogakh Vserossiyskoy perepisi naseleniya 2002. Table 2. As accessed 25 April 2003 from Goskomstat Rossii website, www.gks.ru. 2010: Rosstat, 2010 Census Results, table 1–05.

a more powerful lure than the smaller, distant settlements outside the regional centers and a few other urban centers of growth.

In all but three of the Northern regions, the regional center increased its share of the region's population between 1989 and 2010, even though two-thirds of the regional centers declined in absolute population size themselves. The exceptions were Murmansk (Murmansk Oblast), Dudinka (Taymry Okrug), and Petropavlovsk-Kamchatskiy (Kamchatka Kray). For those regional centers that declined as a share of the regional population, the declines were quite small. For Murmansk and Petropavlovsk-Kamchatskiy, these cities initially declined as a share of the regional population to the 2002 census and then increased by the 2010 census. Dudinka is a somewhat odd case as it is the regional center of the Taymyr Okrug, but not the largest urban settlement located in the region, as nearby Norilsk, which is administratively subordinated to the Krasnoyarsk Kray, is much larger. For those regions where the regional center grew as a share of regional population, the growth was significant. Between the 1989 and 2010 censuses, Nar'yan-Mar (Nenets Okrug), grew from 38 to 52 percent of the regional population, Yakutsk (Yakutia), from 17 to 28 percent, Anadyr' (Chukotka), from 11 to 26 percent, Magadan (Magadan), from 39 to 62 percent, and Yuzhno-Sakhalinsk (Sakhalin), from 22 to 37 percent. Thus, the trend is a universal concentration of the regional populations into the regional centers where there are jobs, access to education and health, consumer goods, and other urban amenities not available in smaller settlements.

Population Change in the Largest Northern Settlements

This section examines population change in the largest settlements in the Russian North, during both the Soviet and post-Soviet periods. There are three broad groups of northern cities. A first group includes cities that are older, have long been part of the Russian/Soviet economy, are located closer to the more densely populated areas of Russia, and were always in the established urban hierarchy of the country, where there was some sort of settlement, albeit small, during the Tsarist period prior to the rapid industrialization and peopling of the North (Map 4.1). This group includes Murmansk (Murmansk), Arkhangel'sk (Arkhangel'sk), Petrozavodsk (Karelia), Severodvinsk (Arkhangel'sk), and Syktyvkar (Komi Republic).

A second group includes those established and rapidly populated during the period of industrialization and urbanization from 1920 to

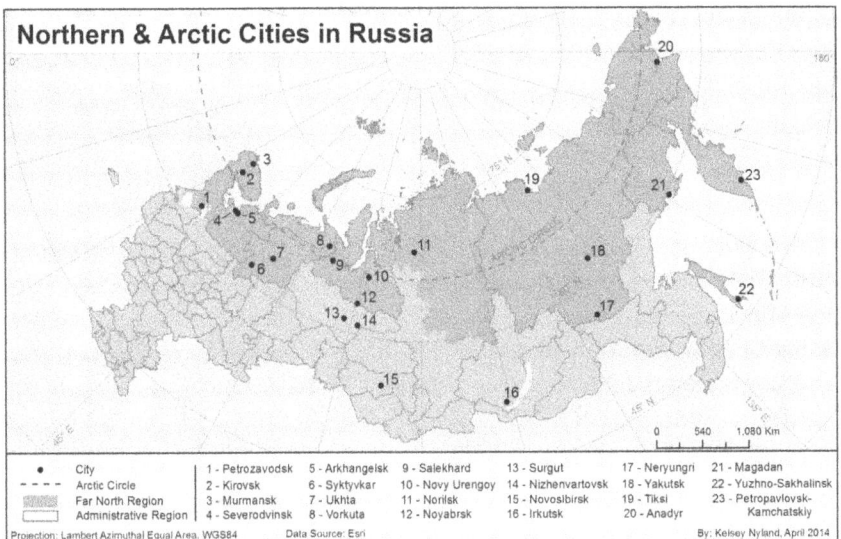

Map 4.1. | *Northern and Arctic Cities*

1960. This group includes Yakutsk (Sakha Republic), Noril'sk (Krasnoyarsk), Yuzhno-Sakhalinsk (Sakhalin), Magadan (Magadan), Vorkuta (Komi Republic), Ukhta (Komi Republic), Apatity (Murmansk), Monchegorsk (Murmansk), and Severomorsk (Murmansk).

A third group includes newer oil and gas towns in west Siberia founded and developed after 1970 including Surgut (Khanty-Mansiy Autonomous Okrug), Nizhnevartovsk (Khanty-Mansiy Autonomous Okrug), Novyy Urengoy (Yamal-Nenets Autonomous Okrug), Nefteyugansk (Khanty-Mansiy Autonomous Okrug), Noyabrsk (Yamal-Nenets Autonomous Okrug), Nyagan (Khanty-Mansiy Autonomous Okrug), and Kogalym (Khanty-Mansiy Autonomous Okrug).

In his classic study of Soviet cities, Chauncy Harris classifies the 304 cities with populations over 50,000 in 1967 into 8 different groups based upon their economic structure (Harris 1970). Of northern cities, Murmansk, Arkhangel'sk, Petrozavodsk, Syktyvkar, Petropavlovsk-Kamchatskiy, Yakutsk, Yuzhno-Sakhalinsk, and Magadan were classified as diversified administrative centers. Severodinsk was classified as an industrial city based on manufacturing. Vorkuta was classified as an industrial city combining mining and manufacturing. The remaining northern cities were unclassified because they had fewer than 50,000 inhabitants in 1967. Those older northern settlements in European Russia were closer and better connected to population centers in central Russia, while those in Siberia and the Far East were distant from other population centers and thus did not benefit from positive agglomeration effects of being close and suffered from the costs of being distant. Thus, most of the larger northern settlements in Russia are quite distant from the core populated areas of the country, and many are either not on the national road or rail system or lack access to both. Many experienced rapid population growth during the Soviet period.

Analysis of population change in the twenty-four largest northern cities (defined here as those with a population of 50,000 or more in 1989) is important for two reasons.[5] One, together they contain nearly half of the population of the North, and, two, many are unique creations of Soviet northern development practices. The largest of the northern cities were the only six cities that lie north of the main inhabited portion of the country, which was most suitable for agriculture in the Soviet Union. This belt stretched from St. Petersburg, southward through Moscow, across Central Russia, and southern Siberia. Of 201 cities of over 100,000 in 1967, only six cities—Petrozavodsk, Murmansk, Arkhangel'sk, Severodinsk, Syktyvkar, Norilsk, and Petropavlosk-Kamchatskiy—were in the North. As Harris states, most cannot be considered regional centers that serve as central places with tributary areas, as most are localized special-function cities. Because so many were creations of a regional development policy that was only possible under the unique conditions of Soviet central planning, it is not surprising that they saw their economies and population shrink considerably after this policy proved unsustainable under the new market conditions in the post-Soviet period.

Collectively, the population of these cities declined by 9 percent between the censuses of 1989 and 2010 (Table 4.3). However, they increased their share of the total population of the North from 43 to 48 percent because there was a larger contraction of economic activity and population in smaller settlements. Based upon their patterns of population change in the post-Soviet period, three different groups of northern cities can be identified (Figure 4.2).

Continued Growth

This group includes Surgut (Khanty-Mansiy), Syktyvkar (Komi Republic), Yakutsk (Sakha Republic), Yuzhno-Sakhalinsk (Sakhalin), Nefteyugansk (Khanty-Mansiy), Noyabrsk (Yamal-Nenets), and Koralym (Khanty-Mansiy). Most of these cities are in the oil and gas regions of West Siberia. Others outside this region that have experienced continued population growth are Syktyvkar, the regional center of the Komi

Table 4.3. | Population Change in Russian North Cities Over 50,000, 1897–2010 (thousands)

City	De facto population								De jure population			Total population (broad definition)			Absolute change, 1989–2010	Percentage change, 1989–2010
	1897	1926	1939	1959	1970	1979	1989	1989	2002	2010	1990	2002	2010		
Murmansk	..	9	119	222	309	381	440	472	336	307	472	336	307	−165	−34.9
Arkhangel'sk	21	77	251	258	343	385	388	417	356	349	428	362	356	−68	−16.3
Petropavlovsk-Kamchatskiy	0	2	35	86	154	215	230	273	198	180	283	198	180	−94	−34.2
Petrozavodsk	13	26	70	135	192	234	271	270	266	262	275	267	265	−8	−2.8
Severodvinsk	..	..	21	79	145	197	249	254	202	192	254	203	194	−62	−24.2
Surgut	..	..	..	..	34	107	248	250	285	307	261	285	307	56	22.6
Nizhnevartovsk	..	..	..	..	16	109	241	245	239	252	247	239	241	7	2.8
Syktyvkar	4	5	24	69	125	171	220	232	230	235	238	246	251	3	1.4
Yakutsk	7	11	53	74	108	152	187	188	211	270	222	246	286	82	43.7
Noril'sk	..	..	14	118	135	180	175	180	135	175	267	222	176	−4	−2.4
Yuzhno-Sakhalinsk	..	..	..	86	106	140	159	156	175	182	171	183	189	25	16.2
Magadan	..	..	27	62	92	121	152	152	99	96	166	106	103	−56	−36.7
Vorkuta	..	..	..	..	..	..	..	115	85	71	217	134	96	−45	−38.8
Ukhta	..	..	3	36	60	87	111	113	103	100	142	127	122	−13	−11.8
Novyy Urengoy	..	..	..	..	..	..	..	95	94	104	107	104	113	9	9.3

(continued)

City	De facto population						De jure population			Total population (broad definition)			Absolute change, 1989–2010	Percentage change, 1989–2010
	1897	1926	1939	1959	1970	1979	1989	2002	2010	1990	2002	2010		
Nefteyugansk	..	..	..	..	..	..	95	108	123	95	108	123	28	29.9
Apatity	..	..	..	..	..	..	88	64	60	114	64	60	-28	-32.2
Noyabrsk	..	..	..	..	..	..	87	96	111	120	103	111	23	26.9
Neryungri	..	..	..	..	..	..	74	66	62	120	90	83	-12	-16.8
Monchegorsk	..	..	..	..	..	..	71	52	45	73	55	48	-25	-36.0
Kotlas	..	..	..	..	..	..	68	61	61	83	74	74	-7	-10.8
Severomorsk	..	..	..	..	..	..	63	55	50	95	75	67	-13	-21.2
Nyagan	..	..	..	..	..	..	55	53	55	..	57	55	0	-0.3
Kogalym	..	..	..	..	..	..	45	55	58	..	56	58	14	30.4

Sources and notes: 1897, 1926, 1939, 1959, 1970, 1979, and 1989: Goskomstat Rossii, Naseleniye Rossii za 100 let (1897–1997): Statisticheksiy sbornik, Moscow: 1998, p. 58. Data refer to the de-facto (*nalichnoye*) population. Missing from the list were data for Vorkuta, Neryingri, and Novy Urengoy and a number of other cities with populations of less than 100,000 in 1989. 1989 and 2002 census figures: Rosstat, VseRoss-iskaya Perepis' naseleniya 2002 goda, Tom 14, Table 4 (www.gks.ru, accessed 15 March 2006). Data are for the de jure (*postoyanoye*) population. 2010: Rosstat, Results of the 2010 All-Russian Population Census, vol. 1, table 11 (http://www.gks.ru/free_doc/new_site/perepis2010/croc/perepis_itogi1612.html). Data are for the de jure (*postoyanoye*) population. Broad definition (includes urban *okrug* or areas subordinated to city, doesn't apply to all): 1990 population figures: Goskomstat RSFSR, Chislennost' naseleniya RSFSR po gorodam, rabochim poselkam i rayonam na 1 yanvarya 1990 g, Moscow: 1990. Data are according to de facto (*nalichnoye*) concept. 2002 census figures: Rosstat, VseRossiskaya Perepis' naseleniya 2002 goda, Tom 1, Table 4 (www.gks.ru accessed 15 March 2007). Data are for the de jure (*postoyanoye*) population. 2010: Rosstat, Results of the 2010 All-Russian Population Census, vol. 1, table 10 (http://www.gks.ru/free_doc/new_site/perepis2010/croc/perepis_itogi1612.html). Data are for the de jure (*postoyanoye*) population.

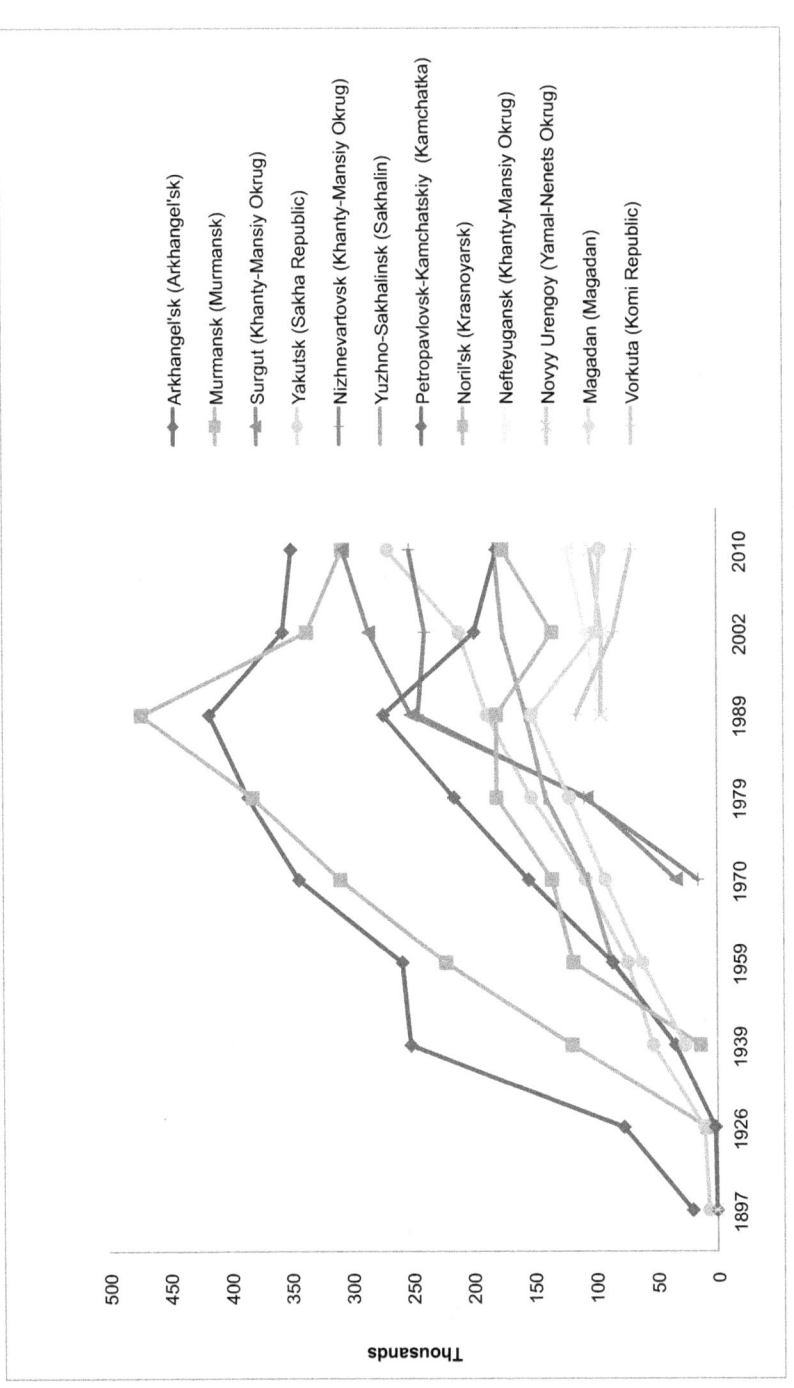

Figure 4.2. | *Population Size in Selected Arctic and Northern Cities, 1897 to 2010*
Source: Russian and Soviet censuses

Republic, which has actually had slow but steady population growth during the post-Soviet period. It is a larger city, well connected to the national transport system and integrated into the national economy. Yakutsk, the regional center of the Sakha Republic, has had the largest population growth of any northern city, growing by 43 percent. This growth was fueled by the region's prosperous diamond industry, the higher natural increase of the younger Sakha population, and also considerable in-migration to the city from regions in Yakutia outside Yakutsk. Yuzhno-Sakhalinsk, the regional center of Sakhalin, which is the center of much of the oil industry on the island, has also experienced continued growth.

Decline and Then Stabilization or Moderate Growth

A diverse group of cities experienced initial population decline and then later stabilization or moderate increase. This group includes Arkhangel'sk (Arkhangel'sk), Petrozavodsk (Karelia), Nizhnevartovsk (Khanty-Mansiy), Noril'sk (Krasnoyarsk), Magadan (Magadan), Novyy Urengoy (Yamal-Nenets), Kotlas (Arkhangel'sk), and Nyagan (Khanty-Mansiy). The population of Arkhangel'sk has fallen by 16 percent since 1989; a portion of the decline can be attributed to natural decrease because of its older age structure. It is a well-connected city to the rest of Russia and has a diverse economy, including the newly formed Northern (Arctic) Federal University. Magadan has had the largest population decline of any city in the North, declining by 37 percent over the period. But after declining from 152,000 in 1989 to 99,000 in 2002, the population seems to have stabilized at 96,000 in 2010. Some cities such as Petrozavodsk (Karelia), Nizhnevartovsk (Khanty-Mansiy), Novyy Urengoy (Yamal-Nenets), and Nyagan (Khanty-Mansiy) have basically had stable populations the entire period.

Continued Decline

Murmansk (Murmansk), Petropavlovsk-Kamchatskiy (Kamchatka), Severodvinsk (Arkhangel'sk), Vorkuta (Komi Republic), Ukhta (Komi Republic), Apatity (Murmansk), Neryungri (Sakha Republic), Monchegorsk (Murmansk), and Severomorsk (Murmansk) all peaked in population size at the time of the 1989 census (or near that year) and have been experiencing continued population decline. This group includes some of the largest cities in the North and others in the European North. Murmansk declined by 35 percent from 472,000 to 307,000. Others in the Murmansk Oblast—Apatity, Monchegorsk, and Severomorsk—also experienced significant economic and population declines. Vorkuta and Ukhta in the Komi Republic were settlements based on the downsizing coal industry. Petropavlovsk-Kamchatskiy, the regional center of the Kamchatka Kray in the Russian Far East, has had considerable increases in the cost of living during the post-Soviet period.

Conclusion

Overall, the population of the Russian Arctic continues to decline, albeit not as steeply as during the 1990s. Natural increase (the difference between births and deaths) is positive in the Arctic, unlike the country as a whole, because of the younger age structure in most Arctic regions. The main driver of population change in the Arctic is migration, which impacts both growing and declining regions. Most of the migration assistance programs that were developed during the 1990s have closed and most migration takes place without state assistance and knowledge. Those who would leave already have and those who will not have decided to continue to make the Arctic or north their home (Mueller and Bradshaw 2006). Evidence of this state of affairs is that for most northern regions, there was a significant downward adjustment of the population estimates following the 2010 census because of unrecorded migration since the 2002 census. In spite of a reputation for immobility, the populations of the Arctic regions are actually quite mobile and are able to adjust to changing circumstances. It would not be a surprise to anybody that Chukotka has the highest rate of outmigration of all Russian regions. It would be a surprise to many that it also has the highest rate of in-migration as well, indicating considerable churning of the population. However, there is strong path dependency to the settlement structure in the Russian Arctic, and it cannot be expected to resemble other Arctic regions with much smaller populations and settlements. With the slowdown in post-Soviet out-migration, it appears as if the population of the Russian North is stabilizing and that the number, size, and distribution of settlements that currently exist will serve as the basis for discussions of urban sustainability in the future.

A recent World Bank report titled *Eurasian Cities: New Realities Along the Silk Road* calls for a rethinking of cities in the region and their role in the economies of these countries (Coulibaly et al. 2012). While the report focuses on the southern regions in Central Russia, many of the lessons apply to the Arctic cities and some even doubly so. The changing demographics and mobility of Eurasian cities is a factor in determining their future. Two decades after the start of economic reforms in Russia, a clear distinction has emerged between a small group of Arctic cities, which are economically booming, and those whose economies have downsized considerably.

According to the report, "the central planners got some things right—easy access to public transportation, district heating networks, almost universal access to water systems, and socially integrated neighborhoods," said Indermit Gill, World Bank Chief Economist for Europe and Central Asia. "But they failed to acknowledge the importance of markets and individual choice in shaping places for people to live in. To become sustainable, Eurasian cities need to find the right balance between markets and institutions." To become catalysts of growth, Eurasian cities need better connectivity, better planning, to become more environmentally friendly, and to mobilize additional financing for these changes. These lessons should selectively be applied to Arctic cities. This chapter has examined the crucial indicator of population change to serve as the basis for other chapters in this volume, which examine other aspects of sustainability in the urban settlements in the Russian Arctic.

Timothy Heleniak is Senior Research Fellow at the Nordic Centre for Spatial Development (Nordregio) and Research Professor of Geography at the George Washington University.

Notes

1. Support for this paper comes from a grant from the National Science Foundation Research Coordination Network–Science Engineering and Education for Sustainability: Building a Research Network for Promoting Arctic Urban Sustainability in Russia (award number 1231294).
2. Though Norilsk is physically located in the former Taimyr Okrug, now Taimyr District, it has always been administratively and statistically a part of the Krasnoyarsk Kray.
3. Data are based on permanent counts of population and do not include shift or fly-in, fly-out workers. There are no comprehensive data on the numbers of people involved in shift work in the Russian Arctic. Other chapters in this volume discuss trends in shift work.
4. Net migration figures calculated via the residual method, where net migration is calculated by subtracting natural increase from total population change, differ slightly.
5. There can be two definitions of the population of an urban settlement. One is a broad definition that could include nearby satellite settlements

and other smaller settlements under the jurisdiction of the core settlement. A second definition is narrower and just encompasses the actual city. For comparability across northern urban settlements, the second narrower definition is used.

References

Arctic Council. 2004. *Arctic Human Development Report.* Akureyri: Stefansson Arctic Institute.
Coulibaly, S., et al. 2012. *Eurasian Cities: New Realities along the Silk Road.* Washington, DC: World Bank.
Dienes, L. 2002. "Reflections on a Geographic Dichotomy," *Eurasian Geography and Economics* 43(6): 443–58.
Harris, C.D. 1970. *Cities of the Soviet Union: Studies in Their Functions, Size, Density, and Growth.* Chicago: Rand McNally and Co.
Heleniak, T. 2009. "The Role of Attachment to Place in Migration Decisions of the Population of the Russian North," *Polar Geography* 32(1–2): 31–60.
Hill, F. & Gaddy, C.G. 2003. *The Siberian Curse: How Communist Planners Left Russia Out in the Cold.* Washington, DC: Brookings Institution Press.
Marinelink.com. 2014. "The Bounds of Arctic Russia Defined." Retrieved 7 May 2014 from http://www.marinelink.com/news/defined-bounds-arctic368543.aspx.
Montes, C. 2014. *American Capitals: A Historical Geography.* Chicago and London: University of Chicago Press.
Mueller, K. and M.J. Bradshaw. 2006. "OPTIMIRUS. Simulating Population Change in the Russian Far East," *European Journal of Population* 22: 105–25.
Rosstat, selected years. *Chislennost' i migratsiya naseleniya Rossiyskoy Federatsii v 20—g. Statisticheskiy byulleten'.* Moscow: Rosstat.

CHAPTER FIVE

Assessing Social Sustainability
Immigration to Russia's Arctic Cities

Marlene Laruelle

Urban sustainability in Arctic regions has many components, including the sustainability of infrastructure in harsh climates, the long-term viability of exploiting resources, and preserving the environment. One of the most understudied sides of sustainability goes beyond these common topics and is related to social changes. This aspect of sustainability is especially important as the global trend toward urbanization is fundamentally changing the environment in which human beings live, including Arctic regions, which also face growing urbanization (Larsen and Fondahl 2014).

Russia represents a unique case study in terms of urban social sustainability. It hosts about 60 percent of the Arctic population (2.9 million people in 2002, 2.3 million in 2015, to which should be added about 10 million inhabitants living in sub-Arctic conditions in Siberia), settled the Arctic region early in historical terms (beginning in the seventeenth century, with the first settlement in Yakutsk as early as 1632), and has built a unique urban fabric there, with half a dozen cities with a population between 100,000 and 300,000 people. Moreover, Russia's economy depends heavily on the Arctic: around 20 percent of the country's GDP and exports are generated north of the Arctic Circle (Medvedev 2008). Because of this combination of historic, demographic, and economic features, Russian Artic cities offer a distinctive place to observe and understand issues related to urban social sustainability.

Urban social sustainability can be defined as a set of elements that includes equity, diversity, social cohesion, quality of life, and governance.[1] Arctic cities have faced a multitude of changes since the collapse of the Soviet Union: reduced "Northern benefits" have become insufficient to compensate for higher prices and the lack of commercial flows from the center (see Chapter 1); industrial production disintegrated rapidly in the early 1990s and has partly revived only in cities with oil and gas or mineral extraction; the decrease in state budget allocations made investments in urban infrastructure a challenge; and municipalities were asked to take the lead in sectors that were previously funded by the central state or the regions.

Linked to these previous trends is the profound change in population and mobility patterns. Although Russia's Arctic regions have been losing population over time, observers have noted a recent contrary tendency in the form of a new wave of in-migrations, which has been going on since the 2000s and especially the second half of the decade. For the population living in Arctic cities, which has grown accustomed to a relatively rigid Soviet-era framework of social stratification, these inflows are perceived as disturbing a social landscape that has already been rendered fragile by the massive outflows of the 1990s. These migrations are not only domestic, with the arrival of Russian citizens, but part of international migration patterns coming from neighboring Commonwealth of Independent States (CIS) countries.

So far research on these migration inflows into Russian Arctic cities remains scarce. This chapter seeks to partially fill the gap by offering an overview of the in-migration patterns. It briefly describes Russia's three Arctic regions and local economic patterns that shape migration flows, investigates the statistical information available to sketch a migration map of Arctic cities, and synthesizes the main sector niches occupied by CIS migrants. It concludes by laying out areas for future research.

Russia's Three Arctic Regions: European, Central, and Eastern

The Russian Arctic is not a unified region in climatic, economic, demographic, or even administrative terms (Stammler-Gossmann 2007; Laruelle 2014: 28–33). Arctic regions are defined as such in some legal documents (for instance, the 2008 Arctic policy that outlines which regions are part of "Russia's Arctic zone"—see Chapter 2), and benefit from specific policies, for instance the delivery of subsided fuel and food (the Northern deliveries) and higher salaries and pensions (the

Northern benefits). However, Russia's Arctic areas do not exist as a unified administrative region. Russia's territorial divisions, a legacy of the Tsarist period perpetuated by the Soviet regime, remain shaped by the conquest of Siberia. The regional divisions run from west to east, gradually stretching from Russia's historical center. The federal districts, which are inclusive of administrative entities with varying statuses, follow this west-east pattern and thus cover both South Siberian and Arctic regions. Russia's administrative divisions also respond to economic rationales: industrial logics often explain the inclusion of some Arctic territories in a region whose economic center is situated further to the south. This regional context influences migration flows, which are largely dependent on local economic dynamics. Russia's Arctic can be delineated into three major regions, schematically definable as the European Arctic, the Central Arctic, and the Eastern Arctic.

The European Arctic, which stretches from Murmansk to Arkhangelsk, is a rather well-defined area. Administratively it is part of the Northwestern Federal District, which includes St. Petersburg, as well as the Barents Sea shipping lanes, which do not belong, legally speaking, to the Northern Sea Route. The region is relatively well connected to both Moscow and St. Petersburg, and the large majority of its population consists of ethnic Russians and Karelians. In terms of domestic development, this region is likely to become part of Russia's "West"—that is, the set of regions whose economies interact and are interlinked with those of its Nordic neighbors (Tykkyläinen and Rautio 2008). Many cross-border projects between Finnish and Russian Karelia, and between Finnish and Norwegian Lapland and the Murmansk region, have taken shape. The Barents Euro-Arctic Transport Area (BEATA), which plans to improve road, air, and rail transport linkages between the Nordic countries and the northwest regions of Russia and to develop joint security projects on external maritime connections, is a good example of these cross-border dynamics (Voronkov 2009). The region has developed infrastructure: it hosts Russia's main naval shipyards (Sevmash and Zvezdochka in Arkhangelsk) and the Northern Fleet; to its west, the Murmansk and Kandalaksha ports are ice-free all year; the fishing industry plays an important role in the local social fabric; the Plesetsk Cosmodrome, Novaya Zemlya, and the Franz Joseph Land archipelago are all key locations for Russia's strategic security across the entire Arctic; and several nuclear power plants as well as mining industries are situated there.

The Central Arctic, which stretches from the Urals to the Taymyr Peninsula, has no unity at the administrative level. It includes the four autonomous districts of Nenets, Yamal-Nenets, Taymyr, and Khanty-Mansiy and the Komi Republic. The Nenets district is attached to the Arkhangelsk region and therefore comprises the easternmost part of the Northwestern Federal District. The Yamal-Nenets and Khanty-Mansiy districts are under the administration of Tyumen, which is itself part of the Ural Federal District. And the Taymyr district was established as part of the Krasnoyarsk region, within the Siberian Federal District. This Arctic area is united by its economic characteristics, particularly its wealth of hydrocarbons and minerals. The energy sector dominates in the western part of this region, while mining prevails more to the east around Norilsk. In both cases, this industrial Arctic is closely related to the more southern region of Tyumen, which has been a center for the extraction of Russian oil since the 1960s, and the region of Krasnoyarsk for the mineral sector. As these extraction sites are the backbone of the Russian economy (Gaddy and Ickes 2005), these regions benefit from large infrastructure plans including: the Belkomur railway project that is planned to connect Finland and Norway with the Trans-Siberian, and be the northernmost railway in the world; an Ob'-Bovanenko line to be built to link the Bovanenko deposit with the extant section of the so-called Transpolar Mainline; and the planned completion of the Salekhard-Igarka railway, an unfinished line dating from the Gulag period. Around Norilsk, the railway connecting mining towns of Talnakh and Kayerkan with the port of Dudinka about 100 kilometers away has been modernized by Norilsk Nickel, as has the port of Dudinka, which offers the largest docking capacity of anywhere along the Northern Sea Route (Humphreys 2011).

The Eastern Arctic includes the republic of Yakutia-Sakha and the Chukotka region. The former is part of the Siberian District, the latter, of the Far Eastern District. Less developed than the two other Arctic regions, with limited transport infrastructure and a whole network of wintertime ice roads (*zimniki*) under threat by permafrost thawing, it is oriented toward the Pacific coast and the dynamism of China further to the south. Yakutia-Sakha hopes to become a hub between the Northern Sea Route and Asia's demand for energy and booming trade, as well as to be more connected with the Irkutsk and Primorye regions (Laruelle 2014). The republic's economy is largely based on the state company Alrosa, the largest diamond producer in the world (Argounova-Law 2004). As far as Chukotka is concerned, it is probably the most marginalized Arctic region, and its economic prospects are the bleakest. Gold extracted from the Kolyma basin, one of the few remaining active industries, operates only in summer. Growing Arctic tourism, coupled with eco-tourism and volcano viewing on the

Kamchatka Peninsula, harbors the potential to revitalize some small settlements, but not the whole region. The main regional economic project remains the transformation of the port of Petropavlovsk-Kamchatskiy in the Avacha Bay into a fishing hub for North Pacific trade (Laruelle 2014), which seems relatively optimistic given the difficulties even Vladivostok has in competing with Chinese, South Korean, and Japanese port activities.

Population flows closely follow these broad economic patterns. The Eastern Arctic continues to lose population, although migration from small towns and villages toward larger towns persists, with Yakutsk attracting the vast majority of those who want to move. The European Arctic faces contradictory flows, with still significant out-migration and a growing in-migration. However, the Central Arctic, which has experienced an industrial revival due to the oil and gas and mineral industries, is most affected by massive inward migration flows.

Population Movement in the Arctic: Not Only Emigration, but Immigration

The collapse of the Soviet system provoked upheavals in migration flows, as Timothy Heleniak has documented in the previous chapter. Between 1987 and 2000, economic output fell by 80 percent in Yakutia-Sakha and Chukotka, some mining centers and industrial settlements were totally abandoned, and several military bases were closed. The downsizing of the Northern Benefits scheme accelerated the exodus of people from the region. The dearth of jobs; dim prospects for young people; exorbitant prices for basic goods; the chronic shortage of heating, gas, and electricity; and declining transportation linkages with the rest of the country pushed millions of Russians to migrate to the European regions of their country (Heleniak 1999; Wites 2006). The majority migrated outside of any state-organized framework. As Heleniak has noted, between 1993 and 2009, the High North "had a population decline of 15.3 percent, consisting of 17.1 percent decline from net out-migration, compensated for by a 1.8 percent increase from the region having more births than deaths as a result of having a younger age structure than the rest of the country" (Heleniak 2010).

Between 1989 and 2006, one out of six people emigrated from the Far North (Heleniak 2008, Heleniak 2009a). Between the censuses of 1989 and 2002, the regions of Magadan and Chukotka lost more than half of their populations, Taymyr and Yamal-Nenets lost 30 percent and 25 percent of theirs respectively, and even Murmansk saw its population fall by more than 20 percent. Yakutia-Sakha escaped relatively unscathed, with a depopulation of only 12 percent (Heleniak 2010). The port towns of Igarka and Tiksi lost about half of their inhabitants between 1987 and 2005, while Dikson lost around 80 percent of its population. In the Magadan region, more than 40 settlements were declared "without inhabitants" in the 2002 census. Ghost towns have grown in number, creating poverty gaps in which the remaining populations do not have enough money to migrate (Andreinko and Guriev 2003). The Far East as a whole lost 17 percent of its population in the space of two decades, declining from 8 million inhabitants in 1990 to 6.4 in 2010 (*Russian Analytical Digest* 2010). The case is similar for the Siberian Federal District, although the decline is less steep.

Internal migration between Arctic regions has been considerable as well (Thompson 2008). Small towns and rural settlements have been abandoned and their inhabitants have moved to larger towns, which are able to provide a wider range of services. But there are also north-south and south-north movements, as the large cities of the Siberian south, such as Krasnoyarsk, attract young people born in the north, who come mainly to attend university before "returning" to their regions of origin (Zamiatina and Yashinskii 2012). Objectively difficult living conditions alone are not enough to make the inhabitants relocate outside the Arctic region. In the first half of the 2000s, the Russian government launched the Northern Restructuring project with a loan from the World Bank. The goal was to assist the voluntary resettlement of Chukotka's non-working population to more southerly towns, but success has been limited and those resettled have experienced difficulties in adapting (Thompson 2002). Indeed, place-specific social capital is not easy to rebuild, and many people have refused to leave the region where they have built their lives, despite the deterioration in living conditions. Arctic identity and feelings of belonging to the region have played an important role in refusals to move (Heleniak 2009b).

However, a more detailed analysis of demographic changes in Russia's Arctic regions yields a less negative and more diverse picture. Just as it was during the Soviet period, the Arctic population is a little bit younger than the national average, both because the oil and gas fields attract youths with a dearth of career opportunities, and because the indigenous peoples have higher birth rates. However, again similar to the Soviet period, life expectancy there is also shorter, both among indigenous peoples and ethnic Russians (Øverland and Blakkisrud 2006). Moreover, towns linked to mineral resource extraction experienced positive migration rates during the 2000s, with Russian citizens and even more foreign migrants coming to enjoy new professional opportunities. It is therefore necessary to distinguish between at least two Arctics: first, regions in crisis that have declining populations, and in which Russians and indigenous populations alike are enduring a degradation of their living conditions; and second, regions in the midst of an economic boom, whose populations are more educated, younger, and more prone to migrate, and which attract an increasing number of foreign migrants. In the latter Arctic zones, migration has more to do with labor market turnover and less to do with a one-way exodus (Heleniak 2009b).

CIS Migrants in Russia's Arctic: A Brief Statistical Overview

Since the fall of the Soviet Union, a series of unprecedented migration flows have rippled across Russia. According to Russian statistics, between 1992 and 2006 3.1 million persons emigrated from and 7.4 million immigrated to Russia, yielding a net positive migration balance of 4.3 million inhabitants (Eberstadt 2010: 153). The figures compiled by the UNDP and the US Census Bureau are higher and show a migration surplus of about 6 million people in the first fifteen years following the Soviet Union's collapse. The majority of Russian emigrants left for Western Europe, Israel, Canada, and the United States, while the majority of immigrants came from among the 25 million ethnic Russians in the CIS, who left their post-Soviet republics to settle in Russia. Today the majority of migrants to Russia's Arctic regions are not native-born Russian citizens, but come from other CIS countries, such as Ukraine, Moldova, the South Caucasus (particularly Azerbaijan), and Central Asia (Tajikistan, Kyrgyzstan, Uzbekistan, and, to a lesser extent, Kazakhstan). The flows of migrants from outside the CIS are smaller. They include migrants from Turkey and Afghanistan, and some specific regional trends, mostly Chinese and, to a lesser extent, Vietnamese and Koreans in the Far East district.

Collecting data on these in-migration flows is challenging. The majority of migrants working in Russia do so as undocumented laborers. Citizens of CIS states (which no longer includes Georgia or Turkmenistan) can enter the Russian Federation without a visa thanks to the 1992 Tashkent Agreement, the freedom of movement treaty that was signed at the fall of the Soviet Union and which is still in force today.

Upon arrival in Russia, emigrants have a month to go to the governing body, the Office of Visas and Registration, to register at a particular location. Emigrants can then obtain a temporary or permanent work permit. Many migrants tend to exceed the legally established time of stay, and ignore or fail to follow the established procedure governing foreigners' labor activity, since Russia's legal and political environment does not encourage the firms that hire them to have them legalized (Reeves 2013a).

Russia's migration policy has been in full swing for years, trying to cope by improving administrative regulation, solving integration issues, and managing economic ups and downs. The years 2013 and 2014 saw eleven Russian federal laws adopted that aim to better regulate migrants and to introduce more severe punishments for violations. The main new rule, which came into force in January 2015, is the requirement to obtain patents for labor migrants from visa-free countries to work in any sector of the economy. The quota/permit, effective since 2007, was abolished, and the patent system became the only document to prove legal employment in Russia (Federal Migration Service 2015).

Collecting precise figures on migration remains largely an impossible mission. Undocumented migrations cannot be precisely calculated. This is the case anywhere in the world, but the Russian situation is specific for several reasons. First, the culture of secrecy pervades sensitive topics such as that of migration. The numbers of migrants are therefore rarely made available to researchers, foreign or Russian, and when they are, it is most often at the level of the entire country, or even that of the major regional entities (the federal districts), but hardly ever at the local level. In addition, the body tasked with migration questions, the Federal Migration Service, is far from a neutral actor. It works in close cooperation with law enforcement agencies, which have every interest in inflating the numbers of migrants in order to win more attention from the political authorities. Migrants' illegality is a profitable business as well: corrupt law enforcement agencies and companies that sell or negotiate false papers have prospered from it (Reeves 2013b).

The researcher must therefore navigate troubled waters with every statistical approach taken. To offer the most comprehensive overview, this chapter will employ four categories of information: the country of origin of Russian citizens, the number of official migrants granted work permits, the unproved projections of undocumented migrants, and the number of internal labor migrations.

Countries of Origin of New Russian Citizens

Between the 2002 and 2010 censuses, population increased in Russia's Central and Southern federal districts, stagnated in the Ural and North-West ones, and fell in the others, albeit at a slower pace relative to 1990: the population in the Far East federal district dropped from 4.6 million to 4.4 million; in the Siberia Federal District, from 13.8 million to 13.5 million; and in the Volga Federal District, from 21.5 million to 20.9 million ("Vot kakie my" 2011).

However, the number of Russian citizens (*Rossiane*) mentioning in their census documents a country of origin outside Russia increased. In 2002, the country counted already 12 million foreign-born individuals, or 8.3 percent of its population, of which 94 percent came from CIS states. The number of new citizens from Ukraine and from Belarus dropped sharply in the 2000s, whereas growth has exploded among the foreign-born population from the southern areas of the former Soviet Union: 40 percent growth for Russian citizens born in Kazakhstan; 70 percent for those from Uzbekistan, Kyrgyzstan, and Azerbaijan; 150 percent for those from Tajikistan; and 219 percent for those from Armenia (Eberstadt 2010: 163) (See Table 5.1). These countries accounted for an intercensus increase in Russia's foreign-born population of about 2.5 million. The booming number of Russian citizens of Azeri and Central Asian origin is indirect documentation of their massive emigration.

CIS Citizens Based in Russia

The statistics for foreigners who arrive in Russia outside the work quota system are, by definition, not reliable. State agencies announce

Table 5.1. | *Foreign-Born Russian Citizens in the 2002 and 2010 Censuses*

Coming from	2002	2010	Percent increase
Azerbaijan	621,000	603,000	−2.9 percent
Armenia	1,130,000	1,182,000	4.6 percent
Kyrgyzstan	31,800	103,000	223.9 percent
Moldova	172,000	156,000	−9.3 percent
Tajikistan	120,000	200,000	66.7 percent
Uzbekistan	122,000	289,000	136.9 percent
Ukraine	2,942,000	1,927,000	−34.5 percent

Source: Rosstat, 2002 and 2010 Censuses

figures when media attention is focused on undocumented migrants (often during and after antimigrant riots), but do not explain the way in which the information is collected and treated. Hence in 2012 the Federal Migration Service declared that there were 12 million foreigners in Russia, blending all categories together (Federal Migration Service 2012). In 2013, the number was raised to 14 million. Of this number, 5 million are just visiting for a few days, while 9 million are staying in the country, with at least 5 million working without a work permit (Egorova 2013). These figures are confusing insofar as it is unclear whether they are dealing with stocks or flows. For instance, many undocumented migrants stay for three months, then exit the country and re-enter it to begin a new three-month stay; thus they are counted twice. Some scholars advance more modest figures of about 3.5 million illegal migrants (Mukomel 2013: 6). Whatever the total number is, Central Asians largely dominate. In 2014 the Federal Migration Service listed about 2.5 million Uzbeks, making them the largest group of migrants, followed by more than 1.5 million Ukrainians, 1 million Tajiks, half a million Kyrgyz, and, around half a million Moldovans, Azeris, and Armenians (see Table 5.2).

Table 5.2. | *Foreign Citizens on the Territory of the Russian Federation, May 2014*

Azerbaijan	600,096
Armenia	491,501
Kyrgyzstan	539,108
Moldova	562,939
Tajikistan	1,137,939
Uzbekistan	2,509,998
Ukraine	1,606,186

Source: Federal Migration Service, http://www.fms.gov.ru/about/statistics/data/details/54891/

Officially Registered Migrants

With only 1.5 million foreign citizens officially allowed to work in Russia in 2013 (see below), one can thus calculate that officially recognized migrants represent between one-third and one-sixth of all foreigners working in Russia. Migrants registered as temporary and permanent residents were part of the quota system the Russian authorities implemented in 2007 in an attempt to reduce illegal flows and provide for companies' labor force requirements. However, the quota policy remained far from satisfying the country's actual migration needs. Since the 2008 economic crisis, the country tried to stimulate the inflow of highly qualified labor migrants, basing it on the American and Canadian models. But the needs for unskilled workers are also immense, and these are what generate the greatest flows of undocumented migrants (Migration Profile: Russia 2013). This quota system was criticized for being simultaneously too rigid and too lax. It was too rigid because firms and regions do not necessarily have the tools to calculate the workers they need for the following year and therefore produce fictitious, inflated numbers to be on the safe side; it was too lax because corruption allows permits to be bought or bypassed with minimal difficulty (Laruelle 2013).

Migrant flows increased in a regular fashion from year to year until the 2014 economic slowdown. Although some sending countries have relatively stable rates (Azerbaijan, Armenia, Moldova, and Ukraine), three countries in Central Asia have seen an exponential increase, especially Uzbekistan. The figures in Table 5.3 combine two types of work permits: the classic work permit granted by companies and the "job-licenses to work for a private person." This latter category grew in 2013 and indicates the extension of the economic model by which private businesspeople directly employ staff. Unlike a company-based work permit, which guarantees minimum standards for salaries, pensions, and health insurance, the job license offers minimal rights to the employed and leaves the door open for many abuses on the part of the employer.

From this table one can see that Uzbek citizens alone count for almost half of the registered foreign citizens allowed to work in Russia, followed by Tajiks and Ukrainians. These numbers give us indirect and undocumented information on the likely similar proportion of nationalities among the nonregistered migrants.

Table 5.3. | *Foreign Citizens Engaged in Legal Labor Activities in the Russian Federation*

	2011 (fourth quarter)	2012 (fourth quarter)	2013 (second quarter)
Total	1,192,000	1,325,000	1,573,000
CIS countries	1,024,000	1,135,000	1,374,000
Azerbaijan	36,000	33,000	33,000
Armenia	53,000	97,000	77,000
Kyrgyzstan	77,000	87,000	93,000
Moldova	55,000	59,000	54,000
Tajikistan	198,000	218,000	271,000
Uzbekistan	476,000	554,000	702,000
Ukraine	117,000	134,000	117,000

Source: Federal Migration Services, quarterly reports

Internal Labor Migrants

Not only is Russia host to CIS migrants, but it also has experienced unprecedented levels of internal migration. Although initially some of this movement may have been related to political tensions in the North Caucasus following the two wars in Chechnya, the flows now are linked to economic conditions and the job market. Following the collapse of the Soviet Union, the number of internal migrants reached 5 million people, only to fall sharply to 1.7 million by 2009. But in the following three years, this figure has doubled to reach 3.8 million in 2012. According to demographers Nikita Mkrtchian and Lilia Karachurina, if one takes into account statistical bias and the fact that some migrations are not captured in statistics, each year between 5 and 6 million Russian citizens move within the country (Mkrtchian and Karachurina 2014). This high number is explained by stark socio-economic inequality between regions, the widening gap between rural and urban areas, and massive departure from the North Caucasus. This unstable and violence-prone region indeed feeds a large portion of internal flows, with a largely net negative migration balance of tens of thousands of people per year (Mkrtchian and Karachurina 2014).

From these four sets of partial data, several major trends can be outlined. "Political" flows related to the collapse of the Soviet Union (the return of ethnic Russians from the former republics, refugees fleeing conflict zones) partially dried up, replaced with migration driven mostly by the needs of the job market, attracting Central Asians, South Caucasians, and North Caucasians. The presence of the first two groups is confirmed by the statistics on the countries of origin of Russia citizens, official migration quotas, and partial information on illegal migrants. The massive exodus of the third group is reflected in internal migration statistics. Being job-related, some of these migrations are driven by employment opportunities from the major extractive industries located in Arctic regions.

Arctic Cities' Power of Attraction

In terms of CIS migrations, Moscow and its surrounding region largely dominate and attract the largest number of migrants due to better prospects for jobs, higher wages, and a well-developed shadow economy (Zaionchkovskaia 2009). After the capital, the most popular destinations are large cities, such as St. Petersburg, Yekaterinburg, and Krasnodar. A third group includes Russia's main industrial centers in Siberia (Tyumen, Surgut, and Khanty-Mansiysk first, then Krasnoyarsk, Novosibirsk, Omsk, and Novokuznetsk), and the main industrial Arctic cities (Norilsk, Salekhard, Yakutsk, Novyy Urengoi, Nar'yan-Mar, and Vorkuta). This migration of workforce age population toward Arctic centers—combined with a higher birth rate for indigenous peoples, and the fact that retirement age people tend to leave for southern regions—give Russia's North a younger demographic structure than the rest of the country.

A similar pattern describes the domestic migration of Russian citizens toward Arctic centers, as demonstrated by Russia's employment statistics. Labor-related migration has been steadily growing over the last decade. In 2005, 600,000 Russian citizens worked outside their place of registration, a figure that reached 4.1 million in 2012, of whom half worked outside their region. Again the region of Tyumen, which includes the Khanty-Mansiy and Yamal-Nenets autonomous districts, had the highest rates after Moscow and Moscow Region, with more than half of workers arriving there coming from the Volga region and 22 percent from other areas of Siberia. Shift work patterns largely explain this trend: major oil and gas projects in the Khanty-Mansiy and Yamal-Nenets districts draw workers from Tatarstan, Bashkotorstan, Udmurtia, and Perm (Mkrtchian and Karachurina 2014; also see Chapter 6 in this volume). As Figure 5.1 shows, cities like Tyumen', Khanty-Mansiysk, and Salekhard that are developing due to oil and gas exploitation, also attract a high level of young migrants from the rural districts surrounding them. Migrants 15–35 years of age are flowing into these cities after leaving other parts of Tyumen Oblast. The only other city in Siberia with a similar profile is Novosibirsk, where the main attraction for young people is the developed university infrastructure rather than jobs in the energy sector (Mkrtchian and Karachurina 2014).

CIS migration and internal migration thus largely overlap in the big oil and gas regions. CIS migrants to the Tyumen region constituted about two-thirds of salaried workers already in the 1990s (Hill and Gaddy 2003: 179). The Gazprom-sponsored Yamal megaproject is expected to require about 50,000 workers, and there were reportedly already nearly 20,000 foreigners working there on infrastructure construction sites at the end of the 2000s ("Investitsii v cheloveka" 2009). Even the state nuclear agency Rosatom has been criticized for employing illegal migrants in its nuclear power plants, as not only do these migrants work in unsafe conditions for low wages, but they are untrained and thereby pose a safety risk at the plants (Ozharovskii 2011). Lastly, the city of Norilsk has by some estimates a population

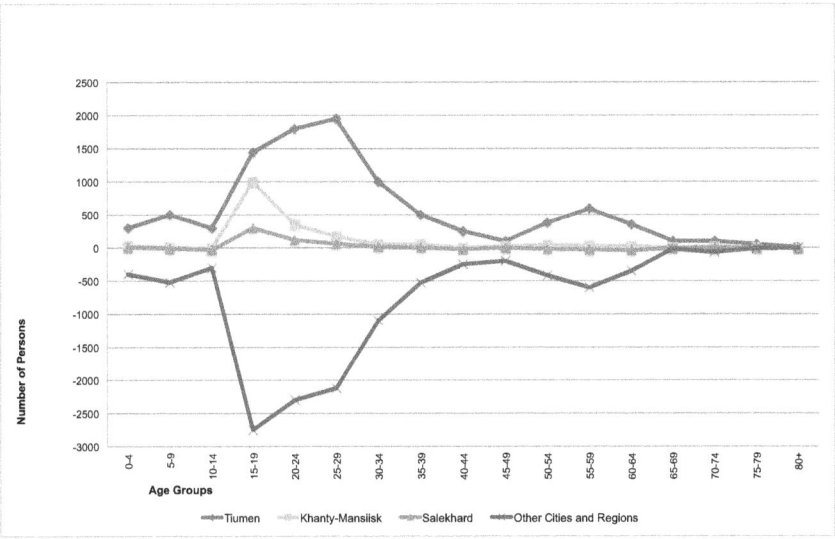

Figure 5.1. | *Net Intra*-oblast *Migration to Tyumen, Khanty-Mansiysk, and Salekhard from Other Cities in 2012 (number of persons by age)*
Source: Mkrtchian and Karachurina 2014.

of 30,000 migrants, mainly from Azerbaijan, Dagestan, and Tajikistan and Kyrgyzstan (Laruelle 2013).

Although statistics and public debates about illegal migrations tend to inflate the numbers and are therefore difficult to use as a basis for analysis, the situation around quotas for legal migrants gives a better idea of the stakes at play. Until the new system put in place across Russia in 2015, firms filed applications for more migrants than the government was likely to assign them. In the Khanty-Mansiy district in 2014, nearly 100 firms requested 19,000 migrants, but the municipality claimed just 6,000 to the Federal Migration Service (Zagumennov 2013). The situation is similar in Surgut, where firms requested 50,000 migrants and the municipality limited the number to just 25,000 in 2012 (Surgutinformtv 2012). The local authorities argue that quotas are never fully used (only between one-third and two-thirds of positions available through the quota system are filled) and thus it is not necessary to increase them. However, these data fail to take into account the highly developed black market for labor. Firms employ many illegal workers, who work for less and require fewer bureaucratic hassles for firm managers. Thus 40,000 migrants came to Surgut in 2012, according to official data from the migration services (Surgutinformtv 2012), a figure closer to the 50,000 that businesses originally requested from the authorities than the 25,000 that the authorities formally allowed. Setting quotas remains a sensitive issue, and some senior regional officials complain that the authorities change their quota limits without notifying firms (Chita.ru 2014).

In terms of nationalities, migrants from Central Asia and Azerbaijan dominate the flows toward Arctic and sub-Arctic cities. These numbers are correlated to the ones available in terms of citizenship by region. Looking at the nationality of Russian citizens,[2] one can observe that between the 2002 and 2010 censuses the number of Russian citizens listing Kyrgyz as their nationality rose 281 percent in the Tyumen region, 243 percent in the Khanty-Mansiy district, 317 percent in the Yamal-Nenets district, and 207 percent in Krasnoyarsk region. The number of Russian citizens listing Tajik nationality follows a similar pattern: 179 percent in the Tyumen region, 173 percent in the Khanty-Mansiy district, 254 percent in the Yamal-Nenets district, and 176 percent in Krasnoyarsk Kray (Rosstat 2002 and 2010). It can be argued that these figures are in themselves unrepresentative, since the scale is only several thousand people. But they reflect the tip of the iceberg, since only a small part of the Central Asian migrants ever obtain a Russian passport. By comparing these figures with, for example, the Azeris and Ukrainians, two national groups traditionally present in the Siberian industrial sectors, one notices that the numbers of Azeris has stagnated, while Ukrainians have declined by about one-third in these same territorial entities (see Figure 5.2), although the numbers began growing again with the refugee flows that followed the 2014 war.

Migrants' Professional Strategies and Their Niches

As in the United States and Europe, migration in Russia has become the main source of labor in some specific economic sectors (Ioffe and Zayonchkovskaya 2011). The extractive industries, construction sites, public service sector (waste management, and road, rail, and water works), and the broader service sector (cooks and waiters in restaurants and cafes, domestic staff such as caretakers, nannies, gardeners, etc.) are all large users of migrant labor. In 2008, CIS migrants in Russia occupied about 40 percent of construction jobs, 19 percent of the trade sector, and 7 percent each in agriculture and production (UNDP 2008: 96). Those present in the Arctic regions can be divided into three broad categories: migrants working for the extractive in-

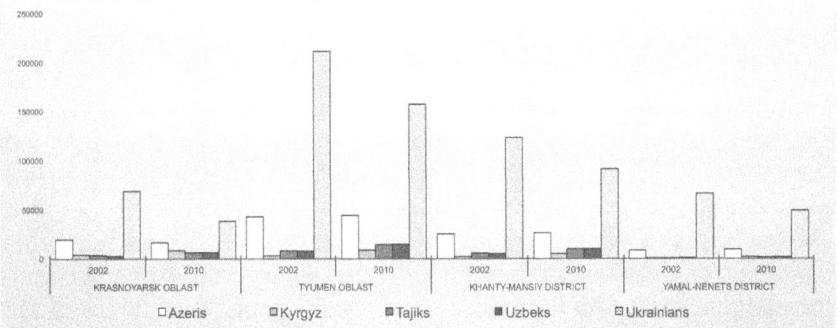

Figure 5.2. *Nationalities of Russian Citizens in Some Arctic Regions*
Source: Rosstat, Russian Population by Nationality by Subjects of the Russian Federation, 2010 Census

dustries, migrants working in the tertiary economy, and migrants occupying professional niches at the bottom of the social ladder. In the following sections, I draw on interviews conducted in Norilsk (unpublished notes, Laruelle 2013 and 2015), to describe these migrants and their sector of work.

"Industrial" Migrants

The first category includes migrants working for oil, gas, and mining companies in Russia's major Arctic and sub-Arctic cities. In the 1990s, companies exploiting the large Siberian oil deposits were the only firms that continued to pay lucrative salaries in the midst of the country's economic meltdown and thus readily attracted Russian citizens from outside the region. In the following decade, this trend slowed down: the rise of the tertiary economy increased the quality of life in the European regions, rendering large and medium-sized industrial Siberian cities less attractive. Consequently, fewer qualified Russian citizens moved to Siberia or the Arctic to take up the offers made by the major companies there, which then looked en masse toward foreign migrants.

Most of these migrants are legal, in the sense that they have work contracts and residency permits. Some arrived via official recruitment mechanisms, at the request of a company; they thus benefit from the legal framework in force. Others were already present, legally or otherwise, and worked in other sectors before managing to get employment with one of the major local companies. The majority of them plan to remain in Russia for the remainder of their professional lives and envisage returning to their home countries for retirement. They often have a professional qualification in hydrocarbons or metallurgy. Many come from Ukraine or Azerbaijan, the two republics that have supplied engineers in the metallurgy, gas, and oil industries since Soviet times. Since the 1990s, there has been a notable inflow of Kazakh engineers, as well. For instance, Norilsk Nickel has brought in several hundred young engineers from Kazakhstan, recruited as they come out of Kazakhstani technical institutes, in order to compensate for the lack of qualified personnel (Laruelle, interviews, 2013). Some migrants who originally had no diploma have succeeded in climbing the ladders of these firms and progressing up the hierarchy. This is the case with many Tajiks. They were the first Central Asian migrants to come to Russia, seeking refuge from the civil war that ravaged their country between 1992 and 1997. Over the next decade, many succeeded in acquiring the skills necessary to hold an engineering position at one of the main Russian companies.

Long-distance commute workers (LDCs) represent a specific case in this first broad category. Shift-work (*vakhtovyi metod*, short-term tours of duty on extraction sites from a base city) is considered to be at the bottom of this industrial social ladder due to its tough climatic and working conditions that require health risks and spending long period away from one's family (Eilmsteiner-Saxinger 2011). Chapter 6 discusses this topic in detail.

"Service Economy" Migrants

The second flow is that of migrants who have secured jobs in niches outside the industrial domain. In the 2000s, the increase in living standards and extensive redistribution of funds from oil revenues redrew the urban geographies of Russia's large industrial towns. What happened in these cities followed, with several years' delay and to a lesser extent, what had already happened in Moscow and the Moscow region. The market economy and especially the boom in tertiary services deeply impacted the cities' geographies. An explosion of construction changed the faces of downtowns and Soviet-era suburbs. New neighborhoods emerged and a symbolic hierarchy between districts arose based on the quality of services offered in terms of small businesses and leisure, modes of life, and consumption.

This second group of migrants is a key actor of this new urban geography. Many of them are middle-aged or older, and migrated to

Russia in the first years following the Soviet Union's collapse or in the 1990s. They are documented or semidocumented (they may have a temporary registration, but not yet a permanent one, for instance), have lived for many years in the same town, and enjoy well-organized community and professional networks. They occupy specific niches related to this bourgeoning tertiary economy: stands at markets selling fresh produce (fruits and vegetables) imported from their home countries; networks of cafes, restaurants, and night clubs; and small construction or transport companies that often secure contracts from larger firms or state bodies. In the Soviet period, South Caucasians (Azeris, Armenians, and Georgians) had traditionally occupied these three sectors. Central Asians were present selling at markets, but not in the entertainment or building sectors. Today, the national dimension of this economic dynamic has been broadly maintained: Azeris are largely dominant in all three sectors.

This group represents an urban bourgeoisie whose members are well established, live in family situations, maintain only periodic relations with their countries of origin, and often have close contacts in local political circles and businesses. They often earned an advanced degree in medicine or education during the Soviet era, but had to change their job profiles after emigrating. They employ other migrants, whether from Azerbaijan, the North Caucasus, or Central Asia, whose status is less established since they arrived later on the job market.

"Bottom of the Ladder" Migrants

The third category of migrants includes all those with precarious statuses, i.e., broadly undocumented. These migrants are overwhelmingly young men who have come alone (either unmarried or leaving their family in their country of origin). They are mobile and move from one city to the next in accordance with work opportunities or their networks. They tend to return at most once a year to their country of origin, often during the winter months, but many had to stay in Russia for years to avoid the risk of not being able to come back. They hold all the unskilled positions that Russian citizens do not want to take: street cleaners (for private firms that have subcontracts with the municipal authorities); workers on building sites (considered a relatively well-paid job but with dangerous working conditions and no guaranteed salary); vendors in street kiosks; and for the luckier ones, taxi and bus drivers, and waiters in cafes and restaurants. Russian citizens tend to disregard these professional niches, deeming the salaries insufficient, working conditions too difficult, and social prestige too low.

This category displays greater national diversity. It includes recently arrived South Caucasians (here also Azerbaijanis) who do not benefit from the networks of the previous generation, Central Asians (mostly Tajiks, Uzbeks, and Kyrgyz), and North Caucasians (mostly Dagestanis). These latter are not migrants, legally speaking, because they are Russian citizens with freedom to travel within their country, but they are socially and culturally marked as "foreign" and find it hard to obtain the requisite registration for work and housing in the city to which they have moved. In addition, the new legislation currently under discussion, which would close the majority of Arctic cities to the internal migration of Russian citizens, would put them in an undocumented situation similar to that of foreign migrants (Arctic-info 2013).

This brief microlevel overview shows glimpses of the situational diversity of CIS migrants in Russia's Arctic cities. Soviet patterns of migration and labor (Ukrainian and Azeri engineers in Siberian industries) are here combined with post-Soviet patterns (young Tajiks and Dagestanis looking for unskilled jobs). The degree of integration of these migrants diverges vastly depending on their status, age, profession, and social capital. Some have become Russian citizens and do not plan to return to their country of origin; some have been settled there for several years, have brought their families with them, and wish to return to their homeland in retirement; some come for a short time and alone to build up some financial capital; some come for a few months to do seasonal work. Some work in abandoned professional niches (street cleaners or bus drivers, for instance); others compete with nationals on the job market (entertainment and leisure sector, for instance).

Conclusion

While in the 1990s the Arctic and sub-Arctic regions were presented as the symbol of Russia's massive depopulation and the abandonment of Siberia by the central government, the picture changed drastically for some regions in the second half of the 2000s. The Central Arctic region, marked by a dense industrial fabric, faced new flows of in-migration due to its need for a cheap workforce, ready to work in difficult conditions. These in-migration flows are local—nearby villages and small towns depopulate in favor of the neighboring industrial city; national—population from the North Caucasus and partly the Volga-

Ural region move to Central Arctic cities; and international—flows from Azerbaijan and Central Asia are massive.

This migration trend will likely increase in the years to come. Regardless of whether all of the Arctic industrial projects currently planned become reality, or whether the demand for labor recedes after infrastructure construction is completed and deposits become operational, the urban fabric has already been profoundly modified by interactions with migrants. Current trends show that migration is decreasingly seasonal and becoming more long-term (Mukomel 2013: 6), meaning that the majority of migrants will not leave even if the economic prospects decline. Moreover, the country's migration policy does not offer solutions, besides massive migration, to one of Russia's major structural problems: its shrinking workforce due to the retirement of the last Soviet generations. Between 2012 and 2017, the working age population will decrease by 1 million persons per year. In 2011 already the Russian Security Council calculated a loss of 10 million workers (out of approximately 70 million currently) by 2025 due to existing demographic trends (Juraev and Bravi 2012).

As in the United States or Europe, migration in Russia is both an opportunity and a challenge for public policies: an obvious economic opportunity for a country lacking a sufficient number of workers, a challenge in terms of social sustainability. Thus far, Russian public policies have had mitigated success in integrating millions of migrants, providing them with legal rights, and protecting them against violence and arbitrary corruption. The country's potential for building a new civic identity has been damaged by the profound social changes that occurred these last two decades, the increase in xenophobia against "southerners"—around 80 percent of Russian citizens would like a visa regime to be introduced for the South Caucasus and Central Asia to limit migration—and the regime's ambiguous and dangerous games in defining national identity.

Cities are the main space where these social and cultural changes linked to migration are lived by the population—even if rural regions are increasingly integrated into these patterns too. Migrants deeply alter the social fabric of Russia's Arctic cities. They contribute to create ethnic districts or at least to give some "ethnic" color to the neighborhoods where they live compactly, develop their own social networks, and open cafes, restaurants, or new places of worship for those who are of Muslim tradition. They also modify the social and cultural urban hierarchies. The definition of who is a newcomer and who is a *korennoe* (local) shifts dramatically. The old pattern of opposing urban Russians to indigenous peoples has been transformed, with the former claiming the privileged status of being "local" and the latter increasingly in competition with CIS labor migrants for penetrating the urban job market and its services.

Research on urban environments in Russia's Arctic regions and their sustainability should therefore include migration as one of the main drivers of changes in the urban landscape, one on which geographers, political scientists, sociologists, and anthropologists can build a multidisciplinary framework of analysis that will include transformations of the urban landscape and of the relationship between the city and its nearby environment; interactions between the private sector and the public authorities, and between the center and the regions; local economic patterns that need workforce and migrants' own professional strategies; new social stratification in terms of capital, values, and identity; and new cultural interactions between "urban dwellers," "indigenous," and "newcomers." Social sustainability is a complex set of interactions that necessitate taking into consideration both material realities and symbolic principles.

Marlene Laruelle is Research Professor of International Affairs at George Washington University.

Notes

1. Three main research institutions work on social sustainability issues: the Oxford Institute for Sustainable Development, the Sustainable Europe Research Institute, and the Institute for Sustainable Futures at the University of Technology Sydney.
2. Even if nationality was deleted from Russian passports in 1997, citizens are still asked about their "nationality" (*natsional'nost'*) dissociated from their citizenship during the census. They are free to make up their answer, can select among a list of more than 100 "nationalities," or can refuse to answer the question.

References

Andreinko, Y. and S. Guriev. 2003. *Determinants of Interregional Mobility in Russia: Evidence from Panel Data.* Moscow: New Economic School.

Arctic-info. 2013. "The Government Will Consider the Bill 'On the Arctic Zone of the Russian Federation' During the Second Quarter." *Arctic-info.com*, 18 March. Retrieved from http://www.arctic-info.com/News/Page/the-government-will-consider-the-bill-on-the-arctic-zone-of-the-russian-federation-during-the-second-quarter-.

Argounova-Law, T. 2004. "Diamonds. A Contested Symbol in the Republic of Sakha (Yakutia)," in E. Kasten (ed.), *Properties of Culture, Culture as Property. Pathways to Reform in Post-Soviet Siberia.* Berlin: Dietrich Reimer Verlag, pp. 257–65.
Chita.ru. 2014. "Galsanov raskritikoval praktiku urezaniia kvot na inostrantsev bez izveshcheniia investorov," Chita.ru, 17 February. Retrieved from http://news.chita.ru/58441/.
Eberstadt, N. 2010. "Russia's Peacetime Demographic Crisis: Dimensions, Causes, Implications," *NBR Report*, May. Retrieved from http://www.nbr.org/publications/element.aspx?id=446.
Egorova, E. 2013. "Illegal Migration in Russia," Russian Council, 6 August. Retrieved from http://russiancouncil.ru/en/inner/?id_4=2195#top.
Eilmsteiner-Saxinger, G. 2011. "We Feed the Nation. Benefits and Challenges of Simultaneous Use of Resident and Long-Distance Commuting Labour in Russia's Northern Hydrocarbon Industry," *Journal of Contemporary Issues in Business & Government* 17(1): 53–67.
Federal Migration Service of the Russian Federation. 2012. *Data on Migration Situation in the Russian Federation.* Retrieved from http://www.fms.gov.ru/about/statistics/data/.
———. 2015. "Press-konferentsiya v edinom migratsionnom tsentre," 16 January. Retrieved from http://www.fms.gov.ru/press/news/news_detail.php?ID=9541.
Gaddy C., and Ickes B.W. 2005. "Resource Rents and the Russian Economy," *Eurasian Geography and Economics* 46(8): 559–83.
Heleniak, T. 1999. "Out-Migration and Depopulation of the Russian North During the 1990s," *Post-Soviet Geography and Economics* 40(3): 281–304.
———. 2008. "Changing Settlement Patterns across the Russian North at the Turn of the Millennium," in M. Tykkyläinen and V. Rautio (eds), *Russia's Northern Regions on the Edge: Communities, Industries and Populations from Murmansk to Magadan.* Helsinki: Aleksanteri Institute, pp. 25–52.
———. 2009a. "Growth Poles and Ghost Towns in the Russian Far North," in E. Wilson Rowe (ed.), *Russia and the North.* Ottawa: University of Ottawa Press, pp. 129–63.
———. 2009b. "The Role of Attachment to Place in Migration Decisions of the Population of the Russian North," *Polar Geography* 32(1–2): 31–60.
———. 2010. "Population Change in the Periphery: Changing Migration Patterns in the Russian North," *Sibirica: Interdisciplinary Journal of Siberian Studies* 9(3): 17–18.
Hill, F., and C. Gaddy. 2003. *The Siberian Curse: How Communist Planners Left Russia Out in the Cold.* Washington, DC: Brookings Institution.
Humphreys, D. 2011. "Challenges of Transformation: The Case of Norilsk Nickel," *Resources Policy* 36(2): 142–48.
"Investitsii v cheloveka." 2009. *Rossiiskie regiony*, no. 4. Retrieved from http://www.gosrf.ru/journal/article/64.
Ioffe, G. and Zh. Zayonchkovskaya. 2011. *Immigration to Russia: Why It Is Inevitable, and How Large It May Have to Be to Provide the Workforce Russia Needs.* Washington, DC: National Council for Eurasian and East European Research.

Juraev, A. and A. Bravi. 2012. "Are Shrinking Quotas in Russia Pushing Migrants into Illegal Work?" *UNDP blog*, 15 March. Retrieved from http://europe andcis.undp.org/blog/2012/03/15/are-shrinking-quotas-in-russia-pushing-migrants-into-illegal-work/.
Larsen, Joan Nymand and Gail Fondahl. *Arctic Human Development Report.* 2014. Copenhagen: Nordisk Ministerråd. Retrieved from http://norden.diva-portal .org/smash/record.jsf?pid=diva2%3A788965&dswid=-1447.
Laruelle, M. 2013 and 2015. Field trip to Norilsk. Interviews with migrants, associations, and city council deputies. July.
———. 2014. *Russia's Arctic Strategies and the Future of the Far North.* New York: M.E. Sharpe.
Medvedev, D. 2008. "Protecting Russia's National Interests in the Arctic," Russian Security Council, 17 September.
Migration Profile: Russia. 2013. *Russia. The Demographic-Economic Framework of Migration. The Legal Framework of Migration. The Socio-Political Framework of Migration.* European University Institute, EU Migration Policy Center.
Mkrtchian N., and L. Karachurina. 2014. "Migratsiia v Rossii. Potoki i tsentry pritiazheniia [Migration in Russia: Flows and centers of attraction]," *Demoscope Weekly*, no. 595–96, 21 April–4 May. Retrieved from http://demoscope.ru/weekly/2014/0595/tema01.php.
Mukomel, V. 2013. *Integratsiia migrantov: Rossiiskaia Federatsiia* [Integration of migrants: Russian Federation]. Saint-Petersburg: European University, KARIM-Vostok Center.
Øverland, I. and H. Blakkisrud. 2006. "The Evolution of Federal Indigenous Policy in the Post-Soviet North," in H. Blakkisrud and G. Hønneland (eds), *Tackling Space. Federal Politics and the Russian North.* Lanham, MD: University Press of America.
Ozharovskii, A. 2011. "Rossiiskie AES stroiat gastartbaitery-nelegaly," *Bellona*, 14 February. Retrieved from http://www.bellona.ru/articles_ru/articles_2011/Rafshan.
Reeves, M. 2013a. "Kak stanoviatsia 'chernym' v Moskve: praktiki vlasti i sushchestvovanie mitrantov v teni zakona [How to become 'black' in Moscow: The practice of the authorities and the existence of migrants in the shadow zone]," in V. Malakho (ed.), *Grazhdanstvo i immigratsiia: kontseptual'noe, istoricheskoe i institutsional'noe izmerenie [Citizenship and immigration: Conceptual, historical, and institutional dimensions].* Moscow: Russian Academy of Sciences/Kanon+, pp. 146–77.
———. 2013b. "Clean Fake: Authenticating Documents and Persons in Migrant Moscow," *American Ethnologist* 40(3): 508–24.
Rosstat. *2002* and *2010 Census*, www.gks.ru.
Russian Analytical Digest. 2010. "Population Statistics of the Russian Far East," no. 82.
Stammler-Gossmann, A. 2007. "Reshaping the North of Russia: Towards a Conception of Space," *Arctic and Antarctic Journal of Circumpolar Sociocultural Issues* 1(1): 53–97.
Surgutinformtv. 2012. "UFMS: nelegal'nykh migrantov v Surgute stalo men'she," Surgutinformtv, 20 August. Retrieved from http://sitv.ru/arhiv/news/social/47773/.

Thompson, N. 2002. "Administrative Resettlement and the Pursuit of Economy: The Case of Chukotka," *Polar Geography* 26(4): 270–88.
Thompson, N. 2008. *Settlers on the Edge: Identity and Modernization on Russia's Arctic Frontier.* Vancouver and Toronto: University of British Columbia Press.
Tykkyläinen, M. and V. Rautio (eds). 2008. *Russia's Northern Regions on the Edge: Communities, Industries, and Populations from Murmansk to Magadan.* Helsinki: Aleksanteri Institute.
UNDP. 2008. *The National Human Development Report. Russian Federation.* Moscow: UNDP.
Voronkov, L.S. 2009. *Geopolitical Dimensions of Transport and Logistics Development in the Barents Euro-Arctic Transport Area (BEATA).* Kirkenes: Barents Institute.
"Vot kakie my—rossiiane. Ob itogakh Vserossiiskoi perepisi naseleniia 2010 g." 2011. *Rossiiskaia gazeta*, 22 December. Retrieved from http://www.rg.ru/2011/12/16/stat.html.
Wites, T. 2006. "Depopulation of the Russian Far East. Magadan Oblast, a Case Study," *Miscellanea Geographica* 12: 185–96.
Zagumennov, A. 2013. "Tochno khvatit? Vlasti Khanty-Mansiyka planiruiut vtroe umen'shit' kvotu na gastarbaiterov [Is there really enough? The authorities of Khanty-Mansiyk plan to cut the quota for guest workers by one half]," *Ura.ru*, 21 June. Retrieved from http://ura.ru/content/khanti/21-06-2013/news/1052160032.html.
Zaionchkovskaia, Zh. (ed.). 2009. *Immigranty v Moskve.* Moscow: Tri kvadrata.
Zamiatina, N. and A. Yashunskii. 2012. "Severa kak zona rosta dlia Rossiiskoi provintsii [The north as a zone of growth for the Russian province]," *Otechestvennye zapiski*, no 5: 225–39.

CHAPTER SIX

The Russian North Connected
The Role of Long-Distance Commute Work for Regional Integration

Gertrude Saxinger, Elena Nuykina, and Elisabeth Öfner

Introduction

For more than six centuries, going back to Yermak's Siberian campaign, Russia has faced the question of how to settle and economically develop the North.[1] With abundant natural resources, this region has long provided the state treasury with revenue. This income became particularly important during the Soviet era, when the extensive exploitation of the mineral resources in the North provided vital support for the industrialization of the country. It was also crucial in ideological terms, helping the USSR remain self-sufficient and independent from Western economies (Hill and Gaddy 2003). Since the beginning of Vladimir Putin's presidency in 2000, the resource-rich territories in Russia's sub-Arctic and Arctic have gained even greater strategic importance: economically, in terms of the international resource trade, and geopolitically, in the context of the competing claims for off-shore resources among the Arctic coastal states (Blakkisrud and Hønneland 2006; Rautio and Tykkyläinen 2008; Rowe 2009; Eilmsteiner-Saxinger 2011). Russia's present economy is based to an unprecedented degree on its mineral resources, which are generally found in the sparsely populated areas of the Far North, including the areas discussed in this chapter: the Yamal-Nenets Autonomous District (YNAO)—Russia's center for gas extraction—and the Republic of Komi, one of the key regions for coal and oil as well as gas extraction.

The Arctic's intersection of economic wealth and geopolitical significance places it at the core of federal policy strategies and discussions on the future prospects of the North. In particular, Russian policy makers must address the question of how to provide the northern extractive sector with an industry-specific labor force while managing the post-Soviet changes in the distribution of the population across the territory.

Political and academic discussions about the future perspectives of the Russian North[2] have centered on two main approaches. The first one, outlined by American scholars such as Hill and Gaddy (2003) and Russian academics such as Pivovarov (2002), supports rescaling the northern territory according to the principles of a free market economy by clustering economic activity and settlements west of the Ural Mountains, close to domestic and international markets. They argued that the North should be developed through the use of long-distance commute (LDC) workers, known in Russia as *vakhtoviki*, and exploited merely as a resource base, but not as a space for permanent habitation. According to this point of view, the main obstacle to Russia's future success is the distribution of labor and capital across its territory, problems that were inherited from the Soviet planning system. Due to its geographical remoteness, extensive distances between settlements, lack of north-north transport connections, and harsh climate conditions, the cost of living in the North is four times higher than in Russia's central and southern regions. This higher cost includes travel expenses, supplies, burdens associated with the challenging climate, such as use of cold-resistant building materials and extra energy consumption, and social costs (Hill and Gaddy 2003: 125). Moreover, the total federal and local government budgets and extrabudgetary funds allocated to support the northern population account for 3 percent of Russia's annual GDP (World Bank 2001). Thus, in order to reduce national expenses and achieve sustainable growth, out-migration of the northern population was encouraged and the closure of economically nonviable and subsidy-demanding northern cities recommended. The industrial development of the North "should be done by reducing the dependency on huge fixed pools of labor and shifting to more technologically intensive methods of extraction and temporary work schemes that do not require a large permanent population or extensive urban infrastructure" (Hill and Gaddy 2003: 213). They stressed that LDC workers should replace permanent settlements in the North.[3]

The second approach, based on a critical response to neoliberal plans for the restructuring of the North, emphasizes the need to overcome the perception of the northern circumpolar region as a "raw materials appendage" to European Russia since its mineral and raw material resources are naturally limited (Melnikova 2006; Voronov 2006). According to this view, Russia's North should keep its population and cities and not become a resource zone to be developed primarily through LDC operations. In contrast to the first approach, which is based on the idea of a minimalistic state and pure economic rationale as the driving force behind northern development, the advocates of the second approach argue in favor of greater state involvement in regional policy in order to keep the overall regional development of the country in balance. They argue that the "contraction of the economic part of the world [oikoumene]" (Pivovarov 2002) under the influence of market economy mechanisms [means] "unavoidably … 'the expulsion of economically developed territory from Russia's economic space,' and also the 'further polarization of regions by level and conditions of life, and even greater contradictions within regions; not only their disintegration, but also the emergence of antagonistic contradictions'" to quote the economist M.K. Bandman in his personal notebook as cited in Melnikova (2006: 38). Instead, northern settlements and their inhabitants should be fully integrated into the spiritual, cultural, military-strategic, industrial, social, and financial space of Russia (cf. Nuykina 2011).

Building on these perspectives, the rationalization of Russia's northern economic geography requires the implementation of resettlement programs that would move socially vulnerable categories of the populations of monoindustrial towns to more temperate zones in the Russian mainland (Heleniak 1999; Thompson 2004; Round 2005; Heleniak 2008; Bolotova and Stammler 2010; Nuykina 2011) and consequently, the downsizing of economically nonviable communities, which are heavily reliant on regional subsidies (Khlinovskaya-Rockhill 2009; Nuykina 2011). On the other hand, communities benefiting from booming economic sectors today should plan for eventual demographic stabilization and block further expansion given the exhaustibility of mineral resources and their dependence on fluctuations in international market prices (Saxinger et al. 2016).

In the long run,[4] demand for additional labor in the hydrocarbon industry is increasing as extraction sites move away from existing mining towns and further north toward the Arctic and into offshore areas. Therefore, the future success of northern hydrocarbon development projects is directly dependent on the further expansion of the use of mobile manpower. Although in the Russian North the number of urban settlements exceeds that of other sub-Arctic regions, the growing labor demand is met exclusively through LDC workers. These workers have become an increasingly important workforce. More and more people become LDC workers and take up a life on the move, requiring them to leave home for long periods, traveling great distances, and regularly subordinating themselves to the closed regime of life in LDC camps (Eilmsteiner-Saxinger 2011, Saxinger 2015, Saxinger 2016c).

This chapter looks back over the last two decades and argues that the two main approaches advocated by politicians, planners, and academics—one proposing a mobile workforce, while the other emphasizes stationary workers in the North—despite being apparently contradictory both function and have developed at the same time, especially in the YNAO and other prosperous industrial regions of the Russian North, but not in disadvantaged Northern regions. However, we see that the generally slow pace of policy implementation in Russia means that there have been few results from these policies. Despite resettlement projects, the North did not turn into a "resource desert" and did not experience social, cultural, or economic disintegration. There are viable towns and settlements, although they are costly and poorly located, according to some analysts (Pivovarov 2002; Hill and Gaddy 2003). The inflow of LDC workers poses new challenges as well as offering opportunities for these existing towns.

The main question we intend to answer is how these two opposing paradigms, neoliberal and conservationist, were reflected in actual practices and to what extent the proposed scenarios of neoliberal planners, as well as those advocated by their critics, have come into being.

Additionally, we strive to answer a second question: How has LDC changed the life of host and home communities and how has it become a means for increased social-spatial proximity and integration of the Russian North and Russian central and southern regions? We put forward the hypothesis that the expansion of the remote northern hydrocarbon industry, with the accompanying employment of mobile workers, can confer benefits on disadvantaged central and southern regions of the Russian Federation and leads consequently to economic growth and the redistribution of wealth.

The answers to these questions are that neither scenario came into being to a substantial extent due to the slow pace of policy implementation on the part of the state over the last two decades. Furthermore, "Putin's Russia" has failed to implement a combination of these policies. Therefore, from a macropolitical point of view, integration of the regions—in terms of fiscal equalization by (among other measures) reducing the supplementary payments for salaries in the North—did not take place (so far). However, in terms of the mental map of individuals who make up the mobile workforce, the northern regions have been brought closer to the central and southern parts of the country through an active practice of social and economic linkages. Therefore, regional integration on a microlevel does successfully exist.

In the following sections, we first provide an overview of LDC workers in Russia. Then we give a brief introduction to the receiving communities in the North (Vorkuta and Novyy Urengoy) and the sending communities in the central regions (Republic of Bashkortostan). Next, we show how LDC impacts these communities. The findings are based on extensive field work: Nuykina in Vorkuta, Saxinger in Novyy Urengoy, and Öfner in the Republic of Bashkortostan. All interviewees' names, except those of experts cited, have been modified to maintain confidentiality.

Both Vorkuta and Novyy Urengoy are the home for intraregional LDC workers as well as being hubs for the distribution of interregional commuters to the workplaces at the remote work sites of the oil and gas industry. Despite these common characteristics, both towns have different histories in terms of the extractive economies: the former is a coal mining region, whereas the latter is characterized by oil and gas production; both faced challenges in terms of economic development over the past two decades. Whereas Vorkuta has experienced the influx of interregional LDC only during the last few years (Nuykina 2013), particularly through its proximity to the extraction fields of the Yamal Peninsula and the construction site of the Bovanenko–Ukhta Pipeline, Novyy Urengoy can be considered a traditional hub for LDC workers since the 1980s (Gustafson 1989: 92, Eilmsteiner-Saxinger 2011).

The Republic of Bashkortostan is located in the southern part of the Ural Mountains. Historically its economy is intimately connected with oil exploration and refining. Since the oil fields in Western Siberia were opened for exploitation in the 1960s, a considerable number of people from the Republic of Bashkortostan are now living in western Siberian cities, or are long-distance commuting back and forth (Stößel 1995; Eilmsteiner-Saxinger 2011; Öfner 2014; Saxinger, Nuykina, and Öfner 2015).

Portrait of the LDC Workers

Presently, no one knows precisely how many LDC workers there are in Russia, since this data is not collected by the government and not officially presented by companies (Eilmsteiner-Saxinger 2011). One illustration of the scale of the phenomenon is the number of LDC in YNAO: their quantity increased more than 2.5 times from 25,500 people in 1991 to 70,200 people in 2008 (Bykov 2011: 50). Forty percent of LDC workers live in the northern regions, while 59 percent permanently reside in the European part of Russia, in the Urals and the southern part of western Siberia. Only 1 percent comes from the former Soviet republics; however, this number constantly grows due to the global economic crisis in 2007/2008 (Bykov 2011: 50).

Long-distance commute work takes two major forms: *interregional* and *intraregional*. The former travel from all over Russia to the oil and gas fields of primarily YNAO and the Khanty-Mansiy Autonomous District (KMAO). The latter refers to residents of northern towns who live near the extraction site, i.e., permanent residents of the North. The distinction between inter- and intraregional long-distance commute work implies differences in terms of shift length and the length of the recreation period: longer rosters, e.g., 30 days on and 30 days off (30/30), 45/30, or 60/30 for interregional and shorter rosters, e.g., 7/7 or 14/14 as well as 30/30 for intraregional long-distance commute workers. However, the shift roster also depends on such factors as profession, function, career level, and the general arrangement in the company (Saxinger 2016a, 2016b, 2016c).

Whereas interregional LDC workers from southern and central parts of the Russian Federation make journeys of up to several thousand kilometers—around 3,000 km to Novyy Urengoy, or up to 9,000 km in the case of Moscow to Sakhalin—the intraregional LDC are permanent residents of base towns near oil and gas fields; however, the latter group may still commute more than several hundred kilometers (see Map 6.1). LDC or fly-in/fly-out (FIFO) are common terms used in mining regions or on off-shore rigs worldwide, as well as on other sites and in sectors where a large labor force is needed in remote areas where local workers are not available (Storey 2010).

Receiving Community: Vorkuta

The town of Vorkuta was established as a coal mining community in the north of the Arctic Circle, 140 km from the Arctic Ocean and 2,268 km northeast of Moscow, in 1938. The discovery of coal resources some years earlier determined the geographical location of the settlement even though it was isolated from other existing population centers and completely disconnected from the central and southern

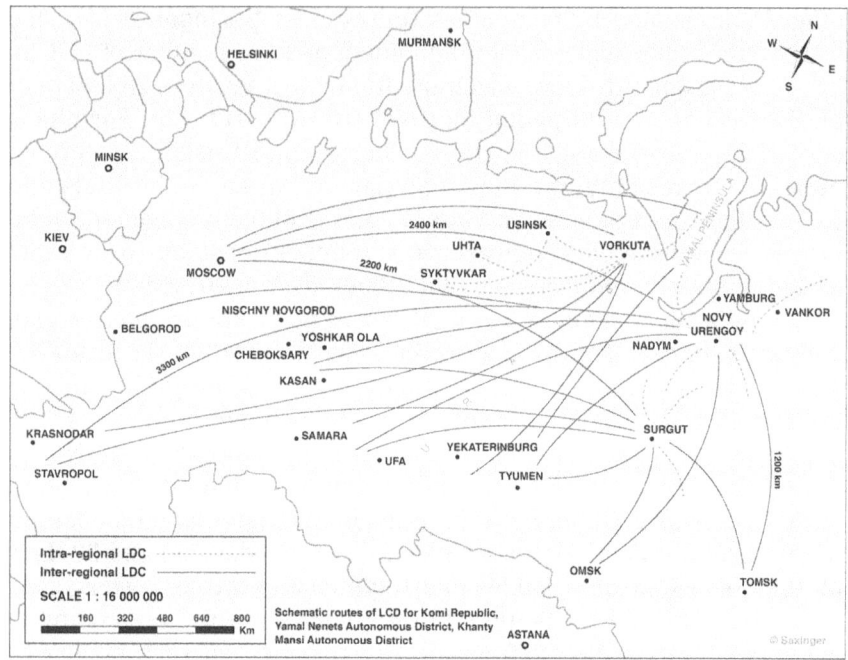

Map 6.1. *Main Routes of Long-Distance Commute Workers in the Northern Urals and Western Siberia*
Source: Saxinger 2016a.

regions of Russia. Coal extraction in Vorkuta started before the first house was built with the prisoners of a Gulag camp purposely expelled to Vorkuta doing the work. Later the town was connected to the central part of Russia via the rail network, which is still the main means of transport for cargo and passenger traffic.

Following the dissolution of the Soviet Union, Vorkuta suffered heavy socioeconomic decline and was advised to close nonviable coal mines, relocate redundant population to warmer regions in central and southern Russia, and to transform into an Arctic hub for LDC workers. At the beginning of the previous decade, the World Bank (2001) developed various programs aimed at making the town economically viable, modern, and less densely populated. However, those projects did not have a significant impact on the area, as many processes developed according to their own agenda (Nuykina 2011).

Vorkuta, together with its satellite suburbs, lost a significant proportion of population, dropping from 212,000 residents in 1986 to 98,500 in 2011, and this natural shrinkage continues (Komistat 2011).

Today, 10,000 locals work in the coal sector. Many of these miners decided to stay in the town despite the restructuring projects and did not follow the mass outmigration of the 1990s (Heleniak 1999). Instead, they began to commute to the industrial sites of the Yamal Peninsula as well as to oil and gas fields near Usinsk and Vyktyl. They also work as intraregional LDC on the giant export gas pipeline project Bovanenkovo–Ukhta. Additionally, Vorkuta receives massive inflows of mobile workers from other areas, coming to the North to construct gas transport pipelines for Gazprom and also to work in the oil and gas and construction sector on the Yamal Peninsula.

Impacts of LDC in Vorkuta

As a central contemporary mining outpost of the Russian circumpolar North, Vorkuta is currently facing new challenges and opportunities related to the inflow of a large mobile workforce coming to and transiting through the city from all over Russia and the countries of the Commonwealth of Independent States. Furthermore, LDC workers come to Vorkuta from different regions, such as the Republic of Bashkortostan, and also from abroad, including from Armenia, Ukraine, Belarus, and Turkey.

Although interregional LDC is a rather recent phenomenon for Vorkuta, its use in the area has increased considerably over the past few years, practically turning the town into a distribution hub for interregional LDC labor. Due to its geographic location, Vorkuta, among other northern urban centers, is regarded as a strategic base for LDC operations mainly on the Yamal Peninsula and the Arctic Urals. According to the internal statistics of the Komiaviatrans airline, there were 1,052 helicopter flights transporting 23,500 workers from Vorkuta to the Yamal Peninsula (one way) in 2011, in addition to the regular train trips.[5] According to another source, there are 6,500 people currently employed in constructing the Bovanenkovo–Vorkuta gas pipeline in addition to 4,000 working on the construction of the Vorkuta–Ukhta pipeline.[6]

Despite the annually growing numbers, LDC workers are not counted in the population statistics of Vorkuta since they are not locally registered. In practice, however, the town is affected by the regular influx of the transient population: some of them stay only for a few hours on the way to the nearby camps or when returning home. Others live in Vorkuta for longer periods of time, depending on their shift and employment agreement.

This tendency can be explained by the lack of local specialists with industry-specific qualifications and experience, better employment opportunities, and a higher income compared to the salaries commute workers can earn in their place of permanent residence. For example, in 2012 a public transport driver earned about US$400 per month in the Astrakhan region, US$800 in the Republic of Bashkortostan, and US$1,500 per shift (one month) working twelve hours a day in the North at a similar job. Since Vorkuta is directly connected to many of the central and southern regions by railroad, allowing for relatively low-cost journeys, an increasing number of workers are traveling to the distant North for work. In addition to the intraregional LDC workforce, Vorkuta supplies domestic labor to Gazprom and other companies, such as the construction company Stroygazkonsulting, involved in constructing the export Bovanenkovo–Ukhta gas pipeline and developing natural gas deposits on the northeast part of the Yamal Peninsula.

The appearance of the new economic players in the region, as well as the influx of LDC workers, influences the economic environment in Vorkuta as well as the everyday life of the community in numerous ways. For instance, demand for housing and accommodation in Vorkuta has increased as gas- and oil-related projects progress. The majority of LDC workers involved in construction live in purpose-built mobile camps outside Vorkuta. Other temporary laborers, especially technical and engineering employees, stay in dormitories and private apartments in the town. The influx of this rotational workforce to Vorkuta has resulted in rising property values over the last few years. In particular, in 2000 a two-bedroom apartment in Vorkuta was worth US$3,000, today, a similar apartment can cost about US$50,000.

The rising housing prices in Vorkuta have two major consequences. First, they raise rents, making life more difficult for those who do not own a private apartment, mostly people with low and middle incomes. On the other hand, higher housing prices provide additional income to the owners since they can rent a flat out to individuals and companies at a higher rate. The rising cost of housing to the point where apartment prices are similar to those in the temperate regions may precipitate outmigration from the North at people's own expense, without reliance on government programs (Nuykina 2011). If they can successfully sell their apartments in Vorkuta, property owners will in turn have the opportunity to purchase a new flat in a small town or village outside the North, which was almost impossible a few years ago.

Profit-seeking citizens are buying up abandoned apartments to convert into simple inexpensive dormitories or hostels for temporary incomers, setting up shops, and opening canteens. New businesses create jobs for some local residents and pay taxes to the municipal budget, which leads to greater diversification and prosperity within the local economy. At the same time, some small-scale enterprises may experience labor shortages, especially in the trade professions. As salaries are relatively higher in the mining sector and also in LDC jobs, local residents do not want to work for lower wages and often chose intraregional commuting as an alternative employment opportunity.

The influx of interregional LDC workers to Vorkuta has also brought a new impetus to the development of the service sector and small businesses in the town. Although in most cases companies provide employees with free meals and accommodation, LDC workers still spend money in local shops on food, mobile communication, laundry services, barbers, cafés, presents for family members, and other goods. Our interviews with LDC workers show that the main expenses when they are on duty include mobile phone top-ups, cigarettes, food, and personal hygiene products. An interregional commuter from Belarus who works at a Baidaratskaya Bay construction site explained that in 2012 he paid around US$170 on average for each such shopping excursion. Alcohol is a minor cost since it is prohibited at work and camp sites (Nuikina 2014).

However, not everybody in the town perceives LDC from an opportunistic point of view, but rather as a threat. Many locals express concerns and fears about the future viability of Vorkuta, mainly in regards to employment and maintaining order in the community. Our respondents in Vorkuta point out that companies prefer to hire workers from other regions as they do not need to pay them according to the northern coefficient (*severniy koeffitsent*), a special compulsory index of salary increases for compensating the costs associated with work and residence in the northern regions, while they are at home in the south during their off-shift. In Vorkuta the additional salary is equal to a 60 percent additional payment. Thus, long-term residents are forced to compete with the cheaper, less demanding, and often more qualified labor streaming in from the southern and central regions.

Another problem repeatedly mentioned by respondents is the negative influence of LDC workers on the social environment in the town and trains, due to drunken and disorderly behavior. Based on local media reports, during the first three months of 2012 there were 154 incidences of disorderly conduct committed by LDC workers in Vorkuta, 125 people were brought to administrative responsibility, and eight commute workers committed crimes (Vorkuta Online 2011). Similarly, the Northern Railway Company reported that in the first quarter of 2012, 66 LDC were taken off the passenger train on the Vorkuta-Moscow line because of inappropriate behavior (Zapolyarka Online 2012a). In addition to that, as in many other cases, the arrival of LDC workers in Vorkuta has been followed by an increase in prostitution and sexually transmitted diseases (Zapolyarka Online 2012b). The latter problem, however, is not unique to Vorkuta as it is experienced in other parts of the globe faced with an influx of temporary male workers (Saxinger and Nuykina 2015).

Regarding the social consequences of LDC, Vorkuta has experienced two sides of the story relating to the issues of integration and isolation of newcomers in the community. The exclusion of commuters from the long-term residents starts on the train, as they travel in separate wagons specially reserved for them by the companies. Moreover, as in the case of Australia, "because of their compressed work schedules and where they live while on site, many FIFO[7] workers do not develop a sense of place and have limited sense of connection to the mining community" (McKenzie 2011: 362). In Vorkuta, those LDC workers who stay in town for the duration of their shift live in separate buildings and camps, although they have unrestricted access to the town and its facilities. This physical separation of the dwellings of temporary workers from the long-term residents, in addition to the cases of inappropriate behavior, has led to the appearance of various rumors and prejudices relating to LDC workers in Vorkuta newspapers and public discourse. On the other hand, recent decisions taken by the municipal administration clearly show official intentions to integrate LDC workers closer to community life by engaging them in city sport and cultural activities. In order to make commute workers more "visible," the city proposed establishing a holiday for gas and oil employees, which would be publicly celebrated in Vorkuta every year in September.

To summarize, on the one hand, the incoming commute workers stimulate diversification and the development of new business activities in Vorkuta, especially in the service sector. Those small and medium companies support large industrial projects with various inputs and required services. The influx also boosts the real estate sector, creates greater diversification of the labor market, and provides additional revenue for the municipal budget. On the other hand, LDC workers are seen as a threat to the existing order in the community and to the employment opportunities of local specialists. Such workers also change the identity of the city as it grows from a monoindustrial mining town into an LDC hub by linking work camps with home regions.

Receiving Community: Novyy Urengoy

In Vorkuta the incoming transient workforce is a recent phenomenon that has altered the shape of city life to the extent that local residents can switch careers and start over as intraregional LDC workers in the petroleum industry and its adjacent sectors. By contrast, the city of Novyy Urengoy in YNAO has had long experience with commute workers. There, LDC is a phenomenon that has shaped the city and the lives of its inhabitants since its very beginning. Working conditions are much more competitive, limiting opportunities for poorly qualified locals.

Novyy Urengoy, often referred to as Russia's "gas capital," was founded in order to extract the natural gas resources of the Urengoy gas field. The first construction work on the gas field's facilities started in the 1970s, and by 1980 Novyy Urengoy had gained district significant city status, becoming a major hub for operations in the area. As more extraction and production sites were developed further away from the city, intraregional LDC had to be introduced, and this became the normal way of life for the inhabitants. Today, as more deposits are discovered, people from Novyy Urengoy also travel to the more distant oil extraction site of Vankor in the Turuchansk district of the Krasnoyarsk Kray, or the gas deposits of Zapolyarnoe or Yamburg and also to Bovanenkovo on the Yamal Peninsula. Both men and women commute to these sites, although the latter work primarily in service and administrative positions as well as in chemical laboratories (Saxinger 2016b).

As discussed in the introduction, the upheavals of the 1990s prompted renewed discussions on how to reshape the Far North, with many advocating shrinking the large cities and relocating the population—primarily the unemployed and pensioners (Nuykina 2011). Waves of similar debates were also ongoing in Novyy Urengoy until the mid-2000s.[8] The neoliberal economic thinking of the 1990s influenced the local approach to demographic policy and city development in general. It was suggested that the interregional LDC workforce, drawing on laborers residing in the central and southern regions of the country, should be increased. However, the idea of shrinking Novyy Urengoy and turning it into "merely" a base town for interregional LDC was met with fierce resistance on the part of the inhabitants as well as the municipal authorities. Today such threats—as these were perceived to be—are a thing of the past (Eilmsteiner-Saxinger 2011).

Novyy Urengoy has stabilized at a size of around 104,000 inhabitants and has the capacity to grow to around 150,000 (for details on this development see Eilmsteiner-Saxinger [2011]). Today monoindustrial petroleum towns like Novyy Urengoy benefit from hosting a qualified and experienced labor force. Qualified specialists further benefit from intraregional LDC, owing to the increasing distances of new extraction sites from cities; as well from north-north interregional LDC. Besides the high salaries in the petroleum sector, special payments (*severnaya nadbavka*) and benefits for LDC are added, such as the so-called northern coefficient, as explained in the section above on Vorkuta. This coefficient is an 80 percent additional payment to the base salary in Novyy Urengoy.

Whereas in Vorkuta, LDC incomers are often perceived as problematic for the city, blamed for driving up housing prices and increasing crime and misbehavior in public places, in the case of Novyy Urengoy, this is not the prevailing view. First, interregional LDC had been established in such a way that the issue of new workers to be settled to Novyy Urengoy on a permanent basis was less troubling; in particular, housing prices are not particularly affected by LDC. If they stay during their stopover, LDC workers are usually accommodated in dormitories or hotels. In general, housing prices in Novyy Urengoy have been at a high level for many years and can be compared to the prices in major cities, such as St. Petersburg and Moscow.

However, the local workforce, and with it the intraregional LDC workers, must compete with the incoming interregional workforce (Eilmsteiner-Saxinger 2011, Saxinger 2016c). As is the case in Vorkuta, in Novyy Urengoy the interregional commuters are generally cheaper than the local workforce due to the northern coefficient, which has to be paid to interregional LDC workers only during their stay in the North. Intraregional workers, in contrast, are entitled to this additional payment all year round. Therefore, an interregional LDC worker on a shift pattern of 30 days on and 30 days off, for example, would incur half of the extra expenses for the added payments compared to a northern worker. Competition among workers is especially intense among low-qualified personnel (Eilmsteiner-Saxinger 2011). Furthermore, interregional workers—especially in the construction sector—are more readily satisfied with lower salaries and more likely to accept degrading working conditions (Saxinger 2016a, 2016b, 2016c). However, general agreements between the companies and the authorities (Komiinform 2012) alleviate this pressure, particularly for more highly qualified personnel. This means that the companies involved would, given the choice between two equally qualified applicants, hire locals in preference to outsiders. Such preferences are of utmost importance to the labor market in the northern cities. However, it is unclear how the situation will develop in the future (Eilmsteiner-Saxinger 2011).

Degrading working conditions are the result of a general restructuring of the petroleum sector in Russia. This industry must compete within an international arena, and is therefore dependent on global markets as well as the dynamics of the price of raw materials. The ongoing outsourcing of nonprofile assets from state enterprises adversely affects the whole region and therefore both the residents as well as the intraregional LDC of Novyy Urengoy (Saxinger 2016c).

Particularly hard hit is the construction sector, which functions within a wide-ranging subcontracting system. In contrast to the Soviet era and the subsequent transition period, Gazprom itself no longer constructs the infrastructure or the plants. Today general contractors, such as the major construction corporation Stroygazkonsulting, outsource to subcontractors who compete and therefore try to lower the costs. In many cases, this competition means lower salaries, while in many reported cases cost savings also occur in terms of workplace security and health care. This practice of continually trying to undercut competitors also affects the inhabitants of Novyy Urengoy, who are used to high salaries. However, living costs remain very high in the North. Therefore, people employed in the subcontracting and sub-subcontracting companies of the construction sector are under increasing financial pressure, especially when it comes to so-called wild commuting (*dikaya vakhta*). This term refers to unregulated commuting and especially to cases when labor laws are bypassed (Saxinger 2015, Saxinger 2016a, 2016b, 2016c). Although there are commissions of various kinds to monitor this issue, the effects have only been curbed, not eradicated (Eilmsteiner-Saxinger 2011).

This competition is particularly relevant when it comes to the employment opportunities of less qualified local inhabitants. As a matter of fact, not everybody is necessarily able to work, or interested in working in the petroleum industry. Furthermore, not all young people manage to pass the exams to enter the various technical educational institutions or can afford the tuition fees. Jobs in other sectors, such as security personnel, taxi and bus drivers, or public service, are paid comparatively poorly and barely cover the high living costs dictated by the extreme northern conditions. These salaries can even be compared to those of the central and southern regions. Such low salaries are common for personnel in the public sector too, including teachers, secretaries in the administration, and similar positions. For instance, the salaries of such jobs are reported to be around US$300 to US$600 per month, and these figures already include the northern coefficient payment. Those who are able become self-employed, go abroad, or work in the informal job market, and some women become sex workers.

As shown above, these are not developments related to the increase of interregional LDC personnel only, but primarily to the changes in the petroleum sector since the dissolution of the Soviet Union, when full employment and social security provided by the state was the norm in the North.

Sending Region: The Republic of Bashkortostan

The Republic of Bashkortostan is one of the main home regions of LDC workers in Russia. In the western part of the republic, oil deposits have been exploited since the 1920s. The skilled labor force for exploration, drilling, extraction, and refining arrived first from the Azerbaijan Soviet Socialist Republic, where oil production began even earlier, and was trained at newly founded vocational educational institutions. Oil extraction peaked in 1967 when the republic's economy was heavily specialized in oil extraction and the refining of petrochemical products. In the 1960s, crude oil and natural gas extraction in western Siberia became more lucrative and many of the workers in the Bashkir ASSR moved to the North. Workers and their families subsequently settled in the newly constructed cities of Novyy Urengoy in YNAO and Nizhnevartovsk, Surgut, and Kogalym in KMAO (Stößel 1995). As the system of LDC was introduced in the late 1970s, specialists were recruited through the labor mobility program *komsomolskaya putevka* (Komsomol, the youth organization of the Communist Party), as well as via university distribution programs (*raspredelenie*). Not least, people were attracted by the high salaries paid in the North and the lucrative benefits. Owing to the strong social ties between the traditional oil regions, including the Republic of Bashkortostan, and the new northern oil region, both the skilled as well as unskilled workforce took up LDC.

These social ties were particularly relevant during the 1990s, when the country was characterized by instability and lack of socioeconomic security. If we look at the example of retired LDC workers, it becomes clear that the first generation of LDC workers benefitted from a good socioeconomic standing during this turbulent decade: One respondent named Salavat grew up in a Bashkir village and received higher education locally and in Moscow. In the oil fields nearby he was employed as a geologist and in 1985 joined a drilling company that operated with LDC workers. He commuted until his retirement in 2010. Although the accommodation in trailer camps was not satisfactory, the high salary compensated for the insufficient living conditions. However, Salavat did not become unemployed, nor did he have to deal with economic difficulties as so many others did in the 1990s: "We were fed and also received small advance payments. Therefore, we were not left without money at that time."

This development on the micro scale is a reflection of the macro scale and reveals the changes in the oil and gas sector when it became the most significant part of the late Soviet economy (Gustafson 1989). People from the Republic of Bashkortostan who were involved in exploration, exploitation, or transportation profited from this development. It also shows that the perception of the oil industry changed. During the Soviet period, employment in the oil sector was considered dirty work. In the 1990s, however, employment in the northern oil and gas industry, with its adjacent construction sectors, for example working on infrastructure, pipelines, or production facilities, became an important opportunity to gain a high and stable income for people throughout the Republic of Bashkortostan. The oil and gas sector became more prestigious and, subsequently, studying at petroleum-related universities became more desirable (Saxinger et al. 2014). Employment in the Northern petroleum sector in the late Soviet period and especially in the oil and gas production sector in the first post-Soviet years facilitated the urbanization of the LDC workers involved. The state enterprises and many of their successors provided housing for the interregional LDC workforce near the LDC transport hubs—such as in Bashkortostan's capital Ufa. This brought about the legacy that LDC workers with a certain work history and stable jobs are located in regional centers.

Furthermore, for white-collar LDC workers or highly skilled professionals, it is possible to earn substantially more income compared to the local average. These people have the tendency to move to or already live in urban areas of Bashkortostan where their spouses find a job easier than in rural homes and children can obtain a better education or can live at home while they are studying at the university.

We found that LDC workers and their families living in urban regional centers of the Republic of Bashkortostan benefit greatly from commuting to the North—workers report earning three to six or more times more than the local average income depending on their qualification and the branch of the petroleum sector they are working in (higher wages in production versus lower wages in construction). Moreover, they show higher social mobility, though the picture is slightly more complex when it comes to LDC workers from rural regions.

Such rural workers face a complex situation when it comes to LDC. On the one hand, LDC is seen as a rather easily accessible income source for a family. It can be attractive because of the higher wage level in the North or because commuting is the sole opportunity to find a job. Employment possibilities are more diversified in the North than in the rural home region. On the other hand, local employment is still preferred over commuting. Such preferences likely have roots in the strong social relatedness to the land that result from Bashkortostan's pastoral history.

LDC is one form of gaining a livelihood, although it is not the most common form of employment and income in the rural areas. Typically, people stay at home and have several sources of income in order to make a living. Making such arrangements requires a widespread social network for gaining access to long- or short-term jobs and to a mutual exchange of help, such as during harvests or when taking goods to markets. Valiakhmetov, Baymurzina, and Lavrenyuk (2015: 114) point out that in 2014 only 29.3 percent of income came through official employment or entrepreneurship; 39.3 percent resulted from nonregistered economic activity and hidden payments. This data shows the precariousness of making a living in rural Bashkortostan.

LDC comes into play here. Income from the North can help families reduce their dependence on other sources of income. If the income from the North is high, some may slowly give up subsistence farming.

In order to make a meaningful income as a LDC worker, rural workers trained in agriculture rather than petroleum first have to acquire the necessary skills and secure access to gatekeepers who can organize a job in the North. Competition among low-skilled workers in the Northern petroleum industry is high, and the strong supply brings about the degradation of wages and working conditions (Saxinger 2016a, 2016b, 2016c).

Despite these issues, the money LDC workers invest in their home communities can be seen as a driving force for local and regional development in general. However, massive mobility to the North could be also an obstacle for further regional development, owing to a shortage of qualified people since working in the north necessarily prevents workers from contributing to their native labor market. Furthermore, Bashkortostan and other sending regions are increasingly dependent on the dynamics of the oil and gas sector, including the boom-bust cycles. Especially in the sector of infrastructure construction for the petroleum industry, the volatility of the oil price brought about significant lay-offs following the 2014 drop.

Regional Linkages

The relationship between the central and southern Russian regions and the North can be understood by examining three types of connections that turn out to have historical roots: first, there are the long-term social ties between those who define themselves as "northerners" and people from the central and southern regions of the country, as we saw in the case of the Republic of Bashkortostan; second, LDC workers in the North have a particular affinity to others of the same ethnic groups in northern and central regions; third, long-term LDC workers perceive the North not as a "cold island" which is "far away," but rather as a close region whose natural resources feed them. For them the mental map has changed and the perception of spatial proximity prevails over distance (Saxinger 2015, 2016a, 2016b).

There has been constant movement and exchange of people between regions such as the Republic of Bashkortostan and the North. Inhabitants from the Bashkir ASSR settled in newly constructed northern cities from the 1960s onwards and have maintained connections to relatives and other social networks in the central regions. As Ledeneva (1998) states, it was advantageous during the Soviet period to have personal contacts with people in the North (or those living in capital cities such as Moscow or St. Petersburg). Through these exchanges social contacts have been strengthened between regions.

Today for example, the lack of universities in the North also strengthens such tight connections: children of northerners move back to the central regions in order to attend higher education institutions. We observed that where relatives live in the central and southern regions plays a significant role in the decision of where the offspring will study.

LDC workers are likely to work together with people from the same region, as in the case of a commuter from the Republic of Bashkortostan who declared that "half of Bashkortostan is in the North." He speaks mostly the Bashkir language at his workplace. Colleagues are in contact during their recreation time in their home region and make agreements to travel back to the North together. Through these tight social networks, common background, and common language, working in the North is perceived as working together with "your own fellows" (Öfner 2014).

Although Vorkuta is located beyond the Arctic Circle, it is linked to the central part of Russia in various ways: via the rail network, people, goods, financial transactions, even through personal histories and memories of the first in-comers brought to Vorkuta by the Stalinist Gulag regime. The extraction of the high-quality coking coal historically connected Vorkuta as a resource base to the metallurgy industries in Cherepovets. The most recent north-south connection includes the export pipeline transporting natural gas from the deposits on the Yamal Peninsula further south. In turn, the central and southern regions supply the town with consumer goods, other materials, and services.

As in Vorkuta, also in the case of Novyy Urengoy, there are four main groups of people who forge links between the North and the central and southern Russian regions. First, there are the long-term residents of the North who came decades ago and perceive themselves as pioneers; some possess a second home in the central or southern regions, such as in the Republic of Bashkortostan. A second group is composed of young people from the North who study or work in the central and southern Russian regions, and pensioners moving back south after retirement. A third group are migrants to the North from various former Soviet states. The fourth group comprise the interregional LDC workers who travel rotationally back and forth.

Building linkages between the North and the central and southern regions is also related to the status of being interregional or intraregional LDC workers. This may change throughout an individual's working life. Permanent residents might switch to an interregional shift work regime, for example, when their already retired spouse and/or their children have moved southwards. Overall, the cases of LDC workers outlined above show that the links between the North and the central and southern Russian regions are both vivid and multidimensional.

Conclusion

In what ways are post-Soviet northern restructuring policies visible in reality today? Over the last two decades or so, efficiency-based neoliberal development plans have been proposed to equalize the Russian regions through "contracting" the North (Pivovarov 2002) and substantially cutting the northern population (Hill and Gaddy 2003; Shaban et al. 2006; World Bank 2006a, 2006b, 2010). Pro-northern and conservationist advocates such as Agranat (1998), Voronov (2006), and Melnikova (2006), have criticized these neoliberal development plans of the 1990s through the mid-2000s for the North because they claim that the neoliberal policies will lead to regional, social, and cultural disintegration among other issues. Our analysis shows in what ways these plans have impacted the contemporary shape of regional integration and relations between the North and the central and southern regions of Russia. In this context, we raise the question of whether, and in which ways, prognoses of disintegration have been realized in practice.

In the beginning of the second decade of the twenty-first century in "Putin's Russia," neither the predictions of the critics nor those of the proponents of efficiency-based neoliberal restructuration policies have come to pass. This moderate outcome is the result of the generally slow pace of implementation of the various and constantly changing development policies for the North.

Although the perks of working in the North are continually being eroded (as we can see from the loss of nonmonetary benefits [*l'goty*]) (Wengle and Rasell 2008), northern monetary benefits and supplement payments, such as the northern coefficient, are still extant. Although northern resettlement programs are under way, their effect today—since the massive uncontrolled outmigration in the 1990s is over—is rather limited in practice and often has controversial consequences (Nuykina 2011; Nuikina 2014).

Nevertheless, the outcomes of prognoses such as those of Melnikova (2006) and Bandman (cited in Melnikova 2006), which claim that levelling the regions and the subsequent abandonment of northern subsidies as well as direct benefit payments to its inhabitants will bring about increased inequality between the regions and for the northerners in particular, can indeed be seen through the general restructuration of the energy sector as described in the examples of Vorkuta and Novyy Urengoy above. This restructuring brought about a substantial decrease in salaries for northerners. Downsizing supplementary payments and benefits for northerners—in order to equalize or integrate Russian regions—would decrease the standard of living even more, since the high living costs in the North are going to prevail.

If integration is understood in the sense of territorial cohesion on the macro level with a balanced distribution of resources in social and economic as well as cultural aspects, this balance and therefore integration have not occurred. If we consider integration from the perspective of contemporary regional development politics on the macro level in Russia, there are also challenges apparent here. Contemporary northern restructuring policies under the Putin regime aim to consolidate the region through shrinking, in such a way as to establish so-called promising economic agglomerations (NIIP Gradostroitelstva 2011), as well as to equalize the regions in fiscal and budget terms.

However, mobile workers from central and southern Russian regions are increasingly benefiting from jobs in the northern petroleum industry and its adjacent sectors, and in addition to this, socioeconomic integration, at least on the micro level, is still taking place—as this process had already begun during the Soviet period, albeit under different mobility policies. The benefits of this integration also go to the sending region since salaries are invested at home and support regional economic development in the more southern parts of Russia.

Today, the integration of the regions can also be seen from another angle: developments in the Russian petroleum industries lead to integration, which means a levelling of increasingly precarious working conditions for both the northern workers as well as those from the south (Eilmsteiner-Saxinger 2011; Saxinger 2016a, 2016b, 2016c). It is therefore important to tackle the problem of partly poor working conditions in the petroleum sector. These are increasingly in violation of contemporary labor law. Furthermore, the growing amount of illegal employment in the context of so-called wild commuting must be addressed. Considering that the petroleum sector forms the backbone of Russia's economy, as well as being its most prosperous sector, it is striking that it is here that the state seems to lack the agency and a clear policy in order to facilitate proper working conditions in a labor system that involves several hundred thousand people (Saxinger 2016a, 2016b, 2016c).

This also applies to the fact that another issue needs to be dealt with: neoliberal approaches in the contemporary petroleum sector—in particular the sub-subcontracting system—are combined with a weak judiciary and corruption in Russia (Saxinger 2016a, 2016b, 2016c). In this regard we agree with Melnikova (2006) that further northern development cannot be left to neoliberal factors only but demands more effective state regulation. In the case of the mobile workforce in the petroleum industry in the North, the state needs to specifically address the bypassing of existing laws and should introduce new standards for the LDC system (Andreev et al. 2009). This basic level of state intervention could prevent LDC, as a method of mobile labor force provision, from becoming increasingly "disavowed" (Bykov 2011).

Today we can see that a combination of labor force provision is prevailing—consisting of the local workforce from viable northern cities as well as of LDC workers from central Russian regions. This combination is a result of failures in resettlement programs on the one hand

and the increase in employment of LDC workers from central Russian regions on the other. The latter development has not been steered by state policy. The increase of the interregional LDC workforce has emerged over time due to the demands of the industry regardless of what preferences the political theories dominant on the various state levels have promoted. Rather, the self-steered development of an increasingly mobile workforce for the North has emerged.

Furthermore, due to this movement of people who are connected in some way with the extractive industries, a tighter relationship between the North and central regions of Russia has been created. This relationship has led to an ever increasing integration of these regions on the micro level in the social and economic context. This integration has come about as a result of the mobile population involved in these spaces and not as a consequence of state policy. For these mobility-oriented people in both sending and receiving regions, social and cultural integration, as well as economic integration, has occurred (Saxinger 2015).

In the North—especially where the extractive industry plays such a crucial role as in Novyy Urengoy and Vorkuta—we see today improved economic well-being compared to that of the average Russian regions (not including exceptional cities such as St. Petersburg or Moscow). Needless to say, Vorkuta is still struggling as a result of the restructuration of its coal mining sector, but can expect a boom period due to the new developments outlined in the case study above. However, two and a half decades after the dissolution of the Soviet Union, the precarious socioeconomic condition in the Russian (rural) regions, such as in the Republic of Bashkortostan, has not yet been effectively addressed. Current equalization policies will more likely bring about the levelling down of northern regions towards the average living standard in Russia, a situation that lacks social and economic security.

Gertrude Saxinger is Assistant Professor in the Department of Social and Cultural Anthropology at the University of Vienna and is affiliated with the Austrian Polar Research Institute (APRI) and the Yukon College in Whitehorse, Canada.

Elena Nuykina is based at the Interregional Expedition Center "Arctic" in Salekhard, Russia.

Elisabeth Öfner is a PhD student in the Department of Social and Cultural Anthropology at the University of Vienna and is affiliated with the Austrian Polar Research Institute (APRI).

Notes

1. Research Project *Lives on the Move*: funded by Austrian Science Fund (*FWF*): [P 22066-G17] (2010-2015). Project partners: Institute for Urban and Regional Research at the Austrian Academy of Sciences (ÖAW) & Department of Geography and Regional Research, University of Vienna, Austria. Project leader: Heinz Fassmann. raumforschung.univie.ac.at/en/research-projects/lives-on-the-move. Further funding: University of Vienna, Austrian Academy of Sciences (ÖAW), Austrian Research Association (ÖFG). We thank the Institute for European, Russian, and Eurasian Studies, George Washington University, for travel grants to attend the meeting on 19 November 2012 in Washington DC. We thank the discussants at this meeting, anonymous reviewers, and our dissertation supervisors (Heinz Fassmann, Peter Schweitzer, Florian Stammler) for their valuable inputs as well as Hannah Gurr for proofreading.
2. For definition of the *Far North* and *Regions Equivalent to the Far North*, see Stammler-Gossmann 2007 and Heleniak 2009.
3. The abbreviation LDC used throughout this article refers in the following to both the workers involved, the long-distance commuters, and to the form of labor organization long-distance commuting, long-distance commute employment, or long-distance commute work. Other terms for this labor organization are fly-in/fly-out (FIFO) and drive-in/drive-out (DIDO).
4. At the moment, however, it is unclear how the massive drop in oil prices since 2014 as well as the international agreement on CO_2 reduction at the United Nations Climate Change Conference in Paris 2015 will impact further developments in the labor market of Arctic resource extraction.
5. Interview with an anonymous expert, Syktyvkar, conducted by Elena Nuykina in February 2012.
6. Interview with an anonymous expert, Vorkuta, conducted by Elena Nuykina in March 2012.
7. FIFO: fly-in/fly-out workers (i.e., LDC workers).
8. Expert interview with V.I. Nuykin, vice-mayor of Novyy Urengoy, Novyy Urengoy, conducted by Saxinger and Aleshkevich, October 2010.

References

Agranat, G.A. 1998. "The Russian North at a Dangerous Crossroad," *Polar Geography* 22: 268–82.

Andreev, O.P., A.K. Arabskiy, V.S. Kramar, and A.N. Silin. 2009. *Sistema menedzhmenta vakhtogo metoda raboty predpriyatiya v usloviyakh kraynego severa* [*The system of management of the fly-in/fly-out method in the work of enterprises in the conditions of the far north*]. Moscow: Nedra.

Blakkisrud, H. and G. Hønneland. 2006. *Tackling Space. Federal Politics and the Russian North*. Lanham, MD: University Press of America.

Bolotova, A. and F. Stammler. 2010. "How the North Became Home. Attachment to Place among Industrial Migrants in Murmansk Region," in C. Southcott and L. Huskey (eds), *Migration in the Circumpolar North: New Concepts and*

Patterns. Edmonton: Canadian Circumpolar Institute Press, University of Alberta, pp. 193–220.

Bykov, V.M. 2011. *Formirovanie konkurentosposobnogo personala v usloviyakh vakhtovogo metoda raboty (Na primere neftegazovoy otrasli)* [Formation of a competitive staff in the conditions of the fly-in/fly-out method of work (the example of the oil and gas sector)]. Yaroslavl: Ministerstvo obrazovaniya nauki RF, mezhdunarodnaya akademiya biznesa i Novykh tekhnologii.

Eilmsteiner-Saxinger, G. 2011 "'We Feed the Nation': Benefits and Challenges of Simultaneous Use of Resident and Long-Distance Commuting Labour in Russia's Northern Hydrocarbon Industry," *Journal of Contemporary Issues in Business & Government* 17(1): 53–67.

Gustafson, T. 1989. *Crisis Amid Plenty: The Politics of Soviet Energy under Brezhnev and Gorbachev, A Rand Corporation Research Study.* Princeton, NJ: Princeton University Press.

Heleniak, T. 1999. "Out-Migration and Depopulation of the Russian North during the 1990s," *Post-Soviet Geography and Economics* 40(3): 155–205.

———. 2008. "Changing Settlement Patterns across the Russian North at the Turn of the Millennium," in V. Rautio and M. Tykkyläinen (eds), *Russia's Northern Regions on the Edge: Communities, Industries, and Populations from Murmansk to Magadan.* Helsinki: Aleksanteri Institute/Aleksanteri-instituutti, pp. 25–52.

Heleniak, T.E. 2009. "Growth Poles and Ghost Towns in the Russian Far North," in E. Wilson Rowe (ed.), *Russia and the North.* Ottawa: University of Ottawa Press, pp. 129–63.

Hill, F. and C. Gaddy. 2003. *The Siberian Curse: How Communist Planners Left Russia Out in the Cold.* Washington DC: Brookings Institution Press.

Khlinovskaya-Rockhill, E. 2009. "The Role of the State in Population Movements in the North: The Case of the Magadan Region, Russian Northeast," *The Role of the State in Population Movements: The Circumpolar North and Other Regions,* MOVE-Conference, Rovaniemi, Finland, 26–28 October.

Komiinform. 2012. "Gazprom i YNAO podpisali soglashenie o sotrudnichestve v 2013 godu [Gazprom and the YNAO signed a cooperation agreement in 2014]." Retrieved 14 December 2015 from www.komiinform.ru/news/94529/#.

Komistat. 2011. *Maloe i srednee predprinimatetel'stvo Respubliki Komi. Statisticheskiy byulleten' nr. 37—92—88/3. Yanvar'—Iyun' 2011g.* [Small and medium business of Komi Republic. Statistical bulletin], Syktyvkar: Federal'naya sluzhba gosudarstvennoy statistiki. Territorial'nyy organ Federal'noy sluzhby gosudarstvennoy statistiki po Respublike Komi.

Ledeneva, A.V. 1998. *Russia's Economy of Favours: Blat, Networking and Informal Change.* Cambridge: Cambridge University Press.

McKenzie, F.H. 2011. "Fly-In Fly-Out: The Challenges of Transient Populations in Rural Landscapes," *Demographic Change in Australia's Rural Landscapes: Implications for Society and the Environment* 12: 353–74.

Melnikova, L.V. 2006. "The Development of Siberia: A Jealous View from Abroad," *Problems of Economic Transition* 48(11): 34–54.

NIIP Gradostroitelstva. 2011. "*Sistema 'Bazovyy gorod—vakhta'* [System 'Base city-fly-in/fly-out]." St. Petersburg: NIIP Gradostroitelstva, unpublished report.

Nuikina, E. 2014. "Making a Viable City: Visions, Strategies and Practices." PhD dissertation, University of Vienna. Retrieved 14 December 2015 from http://othes.univie.ac.at/34120.
Nuykina, E. 2011. "Resettlement from the Russian North: An Analysis of State-Induced Relocation Policy." In Florian Stammler (ed.), *Arctic Centre Reports* 55. Retrieved 14 December 2015 from https://lauda.ulapland.fi/bitstream/handle/10024/59442/AKreport55_electronic110808.pdf?sequence=1.
Nuykina, E. 2013. "Vliyanie vakhtovogo metoda raboty na prinimayushchie goroda Rossiyskogo severa (na primere goroda Vorkuty) [The influence of the fly-in/fly-out method of work on an accepting city of the Russian north]," *Izvestiya Komi Nauchnogo tsentra UrO RAN* 2(14): 107–16.
Öfner, E. 2014. "Russia's Long-Distance Commuters in the Oil and Gas Industry: Social Mobility and Current Developments—An Ethnographic Perspective from the Republic of Bashkortostan," *Journal of Rural and Community Development* 9(1): 41–56.
Pivovarov, J. 2002. "The Contraction of Russia's Economic Ecumena," *Mirovaya Ekonomika i Mezhdunarodnye otnosheniya* 4: 63–69.
Rautio, V. and M. Tykkyläinen, M. (eds). 2008. *Russia's Northern Regions on the Edge: Communities, Industries, and Populations from Murmansk to Magadan.* Helsinki: Aleksanteri Institute/Aleksanteri-instituutti.
Round, J. 2005. "Rescaling Russia's Geography: The Challenge of Depopulating the Northern Periphery," *Europe-Asia Studies* 57(5): 705–27.
Rowe, E.W. 2009. "Policy Aims and Political Realities in the Russian North," in E.W. Rowe (ed.), *Russia and the North.* Ottawa: University of Ottawa Press, pp. 1–15.
Saxinger, G. 2015. "'To You, to Us, to Oil and Gas'—The Symbolic and Socioeconomic Attachment of the Workforce to Oil, Gas and Its Spaces of Extraction in the Yamal-Nenets and Khanty-Mansi Autonomous Districts in Russia," *Fennia-International Journal of Geography* 193(1): 83–98.
Saxinger, G. 2016a. "Lured by Oil and Gas: Labour Mobility, Multi-locality and Negotiating Normality and Extreme in the Russian Far North," *Extractive Industries and Society Journal*, 3: 50–9.
Saxinger, G. 2016b. *Unterwegs. Mobiles Leben in der Erdgas- und Erdölindustrie in Russlands Arktis/Mobil'nyy obraz zhizni vakhtovykh rabochikh neftegazovoy promyshlennosti na Russkom Kraynem Severe/Lives on the Move—Long-Distance Commuting in the Northern Russian Petroleum Industry.* Wien/Weimar/Köln: Böhlau Publishers.
Saxinger, G. 2016c. "Infinite Travel: The Impact of Labor Conditions on Mobility Potential in the Northern Russian Petroleum Industry," in M. Laruelle (ed.) *New Mobilities and Social Changes in Russia's Arctic Regions,* London: Routledge.
Saxinger, G., A. Petrov, V. Kuklina, N. Krasnostanova, and D. Carson. (2016) "Boom Back or Blow Back? Growth Strategies in Mono-Industrial Resource Towns – 'East' & 'West'," in A. Taylor, D. Carson, P. Ensign, L. Huskey, R. Rasmussen, G. Saxinger (eds.) *Settlements at the Edge: Remote human settlements in developed nations,* Edward Elgar Publishing.
Saxinger, G. and E. Nuykina. 2015. "Vakhtoviki i seks-industriya: mify, osvedomlonnost' i deystviye [Fly-in/Fly-out and the sex industry: Myths, awareness,

and action]," in A. YA. Korneeva (ed.), *Psikhologiya ekstremal'nykh professiy: materialy Vserossiyskoy nauchno-prakticheskoy konferentsii 18–19 dekabrya 2014.* Arkhangelsk: Kira, pp. 157–62.

Saxinger, G., E. Nuykina, and E. Öfner. 2015. "Mobil'nost' i migrasiya iz Respubliki Bashkortostan v regiony Kraynego Severa Rossii [Mobility and migration from the Republic of Bashkortostan to the regions of the Russian Far North]," in R.M. Valiachmetova, G.R. Baymurzinoy, and N.M. Lavrenyuk (eds), *Trud, zanyatost' i chelovecheskoe razvitie.* Ufa: Vostochnaya pechat, pp. 225–26.

Saxinger, G., E. Öfner, E. Šakirova, M. Ivanova, M. Yakovlev, and E. Gareev. 2014. "'Ya gotov!' Novoe pokolenie mobil'nykh kadrov v rossiyskoy neftegazovoy promishlennosti ['I am ready!' A new generation of mobile workers in the Russian oil and gas industry]," *Sibirskie istoricheskie issledovaniya* 3: 73–103.

Shaban, R., H. Asaoka, B. Barnes, V. Drebentsov, J. Langenbrunner, Z. Sajaia, J. Stevens, D. Tarr, E. Tesliuc, O. Shabalina, and R. Yemtsov. 2006. *Reducing Poverty Through Growth and Social Policy Reform in Russia.* Washington DC: World Bank.

Stammler-Gossmann, A. 2007. "Reshaping the North of Russia: Towards a Conception of Space," *Arctic and Antarctic Journal of Circumpolar Sociocultural Issues* 1(1): 53–97.

Stößel, E. 1995. *Skizzen aus dem Alltag der Sowjetischen Planwirtschaft: Autobiographische und andere Aufzeichnungen eines rußlanddeutschen Ökonomen aus Baschkirien.* Berlin: Osteuropa-Institut.

Storey, K. 2010. "Fly-in/Fly-out: Implications for Community Sustainability," *Sustainability* 2: 1161–81.

Thompson, N. 2004. "Migration and Resettlement in Chukotka: A Research Note," *Eurasian Geography and Economics* 45(1): 73–81.

Valiakhmetov, R.M., G.R. Baymurzina, and N.M. Lavrenyuk. 2015. *Trud, zanyatost' i chelovecheskoe razvitie: doklad o raszvitii chelovechskogo potentsiala v Respublike Bashkortostan* [Work, employment, and human development: A report on the development of human potential in the Republic of Bashkortostan]. Ufa: Vostochnoy pechat.

Vorkuta Online. 2011. "Vakhtovikov izoliruyut ot obychnikh passazhirov [Fly-in/Fly-out workers are isolated from ordinary passengers]." Retrieved 14 December 2015 from http://vorkuta-online.ru/index.php/2010-12-14-18-50-34/2458-2012-04-18-06-01-16.html#comment-242.

Voronov, I.P. 2006. "Siberians Sitting on Their Suitcases," *Problems of Economic Transition* 48(11): 55–70.

Wengle, S. and M. Rasell. 2008. "The Monetisation of L'goty: Changing Patterns of Welfare Politics and Provision in Russia," *Europe-Asia Studies* 60(5): 739–56.

World Bank. 2001. *Project Appraisal Document on a Proposed Loan in the Amount of US$80 Million Equivalent to the Russian Federation for a Northern Restructuring Project.* Washington, DC: World Bank Human Development Sector Unit.

———. 2006a. *Analytical Report on the Lessons and Expansion of the Northern Restructuring Pilot Project.* Moscow: Foundation for Enterprise Restructuring and Financial Institution Development.

———. 2006b. *Quantitative Report on Pilot Project of Social Restructuring of the Far Northern Territories.* Moscow: Foundation for Enterprise Restructuring and Financial Institution Development.

———. 2010. *Russian Federation—Northern Restructuring Project.* Washington DC: World Bank.

Zapolyarka Online. 2012a. "Povedenie vakhtovikov "Gazproma" stalo predmetom razbiratel'stv v administratsii Vorkuty [The behavior of Gazprom fly-in/fly out workers has become an issue in the Vorkuta administration]." Retrieved 14 December 2015 from http://заполярка-онлайн.рф/vorkuta/povedenie-vax tovikov-gazproma-stalo-predmetom-razbiratelstv-v-administracii-vorkuty.html.

———. 2012b. "Anatoliy Puro ustroil gazovikam 'razbor poletov' razdav i knuty i pryaniki [Anatoliy Puro arranged a debriefing for gas workers, distributing carrots and sticks]." Retrieved 14 December 2015 from http://заполярка-онлайн.рф/vorkuta/anatolij-puro-ustroil-gazovikam-razbor-poletov-razdav-i-knuty-i-pryaniki.html.

Section III

Climate Change

CHAPTER SEVEN

Cities of the Russian North in the Context of Climate Change

Oleg Anisimov and Vasily Kokorev

Introduction

In addressing Arctic urban sustainability, one has to deal with the complex interplay of multiple factors, such as governance and economic development, demography and migration, environmental changes and land use, changes in the ecosystems and their services, and climate change.[1] While climate change can be seen as a factor that exacerbates existing vulnerabilities to other stressors, changes in temperatures, precipitation, snow accumulation, river and lake ice, and hydrological conditions also have direct implications for Northern cities. Climate change leads to a reduction in the demand for heating energy, on one hand, and heightens concerns about the fate of the infrastructure built upon thawing permafrost, on the other. Changes in snowfall are particularly important and have direct implications for the urban economy, because, together with heating costs, expenses for snow removal from streets, airport runways, roofs, and ventilation spaces underneath buildings standing on pile foundations built upon permafrost constitute the bulk of a city's maintenance budget during the long cold period of the year. Many cities are located in river valleys and are prone to floods that lead to enormous economic losses, injuries, and in some cases human deaths. The severity of the northern climate has a direct impact on the regional migration of labor. Climate could thus potentially be viewed as an inexhaustible public resource that creates opportunities for sustainable urban development (Simpson 2009). Long-term trends show that climate as a resource is, in fact, becoming more readily available in the Russian North, notwithstanding the general perception that globally climate change is one of the greatest challenges facing humanity in the twenty-first century.

Like the rest of the world, Russian society is divided between those who believe that climate change will have a major impact on the planet and those who doubt such predictions: alarmists and skeptics as the media often describes them. Public opinion polling shows that there is no consensus and identifies various cleavages in Russian society. According to a 2008 survey of more than 1,000 people from different regions of Russia (WCIOM 2008), 48 percent of people have heard something about climate change, while 8 percent do not know what it is; 51 percent believe that global climate change has already started (compared to 48 percent in 2007); 57 percent associate climate change with human influences, whereas 29 percent believe it is caused by natural factors; 50 percent (up from 45 percent in 2007) believe that climate change may have catastrophic impacts, while 27 percent think it will not have any serious environmental consequences, and another 7 percent believe that some regions could benefit from the changing climate.

Researchers conducted another survey in September 2010, immediately following an unusually long heat wave in central Russia, which sparked numerous forest fires over large swathes of land (WCIOM 2010). Data from this survey indicate that most people (57 percent) associated the increased number of fires with inappropriate management and the so-called human factor, which in Russia generally indicates a lack of discipline, responsibility, and proficiency, as well as the unwillingness and/or inability of individuals and officials to follow established regulations. Only 34 percent of those polled equated the increased frequency of fires to global climate change.

Andrei Illarionov, founder and director of the Russian Institute of Economic Analysis (IEA), best exemplifies the climate skeptics who are most prominent in the political arena. From 2001 to 2005, he served as economic adviser to President Vladimir Putin. As a political and social leader, Illarionov is known in Russia for his public statements, many of which tend towards sensationalism. His views on climate change are in conflict with the accepted scientific consensus, as can be seen in numerous media interviews and in distilled form in the paper "How to Spin Warming: The Case of Russia."[2] Interestingly, the paper was released on 16 December 2009, at the peak of the "Climategate" campaign in which a hacker gained access to and published some e-mails exchanged by climate scientists on the eve of the Copenhagen summit on climate change. Climate change skeptics claimed that the e-mails demonstrated that climate change was a hoax. In his paper, Illarionov accused the Intergovernmental Panel on Climate Change (IPCC) of selective use of data from approximately 25 percent of Russia's weather stations. According to the IEA, the selected stations overestimate the rate of warming in Russia, questioning the credibility of the IPCC findings. His paper provoked extensive discussions in the media and had a pronounced societal impact, particularly on those who claim that climate change is simply a matter of belief.

Meanwhile, the majority of the scientific community addresses the problem from the other end, considering climate change as a matter of fact. The idea for global climate change and its scientific basis can be traced back to the 1960s, when Russian professor Mikhail Budyko published his papers on what is now called "geoengineering" (Budyko 1962). In that work, Budyko proposed that it was possible to alter the global climate through the mechanism of changing the albedo of polar ice by dispersing soot. In the following years he developed the first numerical climate model, which was published in 1969. For the first time ever, he linked the anthropogenic combustion of fossil fuels with the growing atmospheric concentration of CO_2 and global air temperatures. Budyko lived long enough (he passed away in 2001 at the age of 80) to witness the end-of-century patterns of a changing climate, which he had successfully predicted in 1969, at least in general terms sufficient for large-scale adaptation planning. In the 1970s, he published articles describing his ideas in popular magazines, which is why both the general public and authorities in Russia were relatively well aware of climate science much earlier than their counterparts in the West, where the first publications and public debates started later. In the mid-1980s, well before the IPCC came into existence, Soviet authorities charged Budyko's department of climatology at the State Hydrological Institute in St. Petersburg with the task of assessing the implications of climate change and contracting sea ice for Arctic marine navigation. Their particular interest was in strategic planning to develop the icebreaker fleet with an optimal split between nuclear and conventional vessels to serve the transportation needs of the Northern Sea Route, as well as those on Siberian rivers. Research along these lines was unfinished at the time the USSR collapsed, and the subsequent political and economic breakdown of the country led to its early cessation.

In the decades following the 1960s, climatology evolved from a descriptive and mostly empirical discipline to a highly comprehensive science with a strong computational component. Currently, it operates with fundamental and complex physical equations, and employs sophisticated three dimensional thermodynamic modeling. Models account for numerous interacting components of the climatic system, and ultimately generate millions of digital parameters in order to draw climatic pathways from the past to the future. The amount of data generated by climate models is enormous, and not surprisingly, the general public as well as scientists outside the climatological community often lack the scientific background to engage with climatological models.

The purpose of this chapter is to provide a scientifically comprehensive yet nontechnical overview of the current and projected climate in the northern Russian regions. It places particular emphasis on the climatic parameters that have potential implications for cities in the Far North. As discussed in the introduction to the book, we intentionally define the boundary of what we call the "North" loosely to include mountainous regions occupied by permafrost in southern Siberia, Altai, and Kamchatka. Largely in response to the criticism expressed in Illarionov's papers, we pay special attention to the evaluation of observation certainty and the identification of gaps in the climate data. This analysis is based on Russian hydrometeorological service data and the results from the latest generation of the most sophisticated Coupled Model Intercomparison Project Phase 5 (CMIP5) climate models. To minimize the subjective component, standard IPCC methodology has been employed. The analysis begins with a discussion of regional features in the temperature and precipitation patterns using observations from the twentieth and early twenty-first centuries, an evaluation of predictive climate models in a regional context, and construction of comprehensive climatic projections for the future on the basis of the best models. Subsequently, key regional concerns and opportunities associated with climatic changes are identified, concluding with an assessment of the direct and indirect impacts climate change may have on the cities of the Russian North.

Regional Climatic Changes in Russia in the Context of Global Warming

Global climate change in the twentieth century was characterized by warming through all seasons, changes in seasonality, an increased frequency of extreme weather events, and increasing durations of periods with the temperatures above or below prescribed thresholds.

Many regions, including Russia, demonstrated discernible changes in the intensity and frequency of precipitation. Changes were uniform neither across space nor over seasons; in Russia, however, many of the changes exceeded global means.

Observational records contain two periods of pronounced hemispheric-scale warming in the early and late twentieth century (Hansen et al. 2010). The first period of warming started in the 1920s and lasted for nearly two decades. The second warming began in the late twentieth century and continues through the present, with unequivocal evidence demonstrating a combination of natural and anthropogenic factors driving these changes (Solomon et al. 2007). In this section, we compare variations in the global temperature and regional climate in Russia.

The problem is threefold: First, it is essential to establish whether records at Russian weather stations point to discernible signals of climate change, and if so, what type of change it represents. The second task is to develop a regionalized analysis to characterize the pattern of modern changes. Last, we need to analyze regional-mean trends, compare them with the changes at a global level, and build regional projections for the future.

We address the first task—determining whether the information collected by Russian weather stations point to climate change—through the analysis of temperature data. The rationale behind this choice is that temperature is the only climatic parameter that responds directly to changes in the radiative forcing (the difference between sunlight received by the earth and radiated back to space) produced by greenhouse gases through well-understood physical mechanisms. Temperature is thus the ultimate factor governing the cascade of changes in atmospheric and oceanic parameters and processes, which in their totality constitute climate change. We tested three types of statistical models to select the one that provides the best fit to the observations:

- A stationary time series model that rejects the concept of climate change, suggesting that observations correspond to the stationary regime characterized by the natural variability and the same mean value
- A time series with linear trends
- A time series with stepwise changes from one stationary regime to another

We applied each of the three types of model to century-scale temperature data from Russian stations and used the standard deviation of the simulated time series from observations as a metric of the model fit. The stationary model demonstrated the worst fit; on average it had an 11.2 percent higher deviation than the linear trend model and 10.3 percent higher than the stepwise-change model, while at many stations the difference in deviation was more than 20 percent. As for the other two models, the linear trend model consistently demonstrated a better fit with slightly lower deviation. We thus conclude that there is observational evidence of a changing climate in Russia, which could be best approximated by a linear rise in temperature over time.

To address the second task, describing the pattern of change across regions, we combined the records from individual stations into groups and developed a climatic regionalization based on the coherence of the temperature variations. Averaging data over the coherent regions allows the signals of climate change to be highlighted by minimizing the stochastic component, which is present in the individual station records.

We tested several classifications consisting of different numbers of regions using data for different periods. The regional delineations were based on the federal administrative units of Russia, which were in some cases divided into smaller subunits to achieve homogeneity of bioclimatic and topographic conditions. The optimal climatic classification for the modern period is shown in Map 7.1 and consists of

Map 7.1. | *Location of Weather Stations, the Main Population Centers in the Russian North (defined here as the area within permafrost boundaries and/or northward of 60°N), and the Southern Permafrost Boundary; Map Partitioned into Regions with Coherent Temperature Changes in the Period 1970–2010*

seventeen regions, of which 1–14 are in the Russian Federation. In the context of the present study, regions 1, 7, and 9–13 located within the Russian permafrost area and/or northward of 60°N are of prime interest. Map 7.1 also shows the location of the permafrost boundary, weather stations, and main population centers in Northern Russia.

The coherence of the temperature changes in Russia (regions 1–14) following the early-century peak is illustrated in Figure 7.1. The upper left plot was constructed using the CRUTEM3 data and shows the hemispheric-mean mean annual air temperature (MAAT) smoothed by an eleven-year running window. Curves in plots 1–14 represent MAAT changes at individual stations (grey lines) and the regional-mean (black line) smoothed by an eleven-year running window. Although not shown here, similar plots were constructed for precipitation and other temperature characteristics, such as seasonal temperatures and temperature sums above and below prescribed thresholds that have direct implications for city management. Table 7.1 provides a sum-

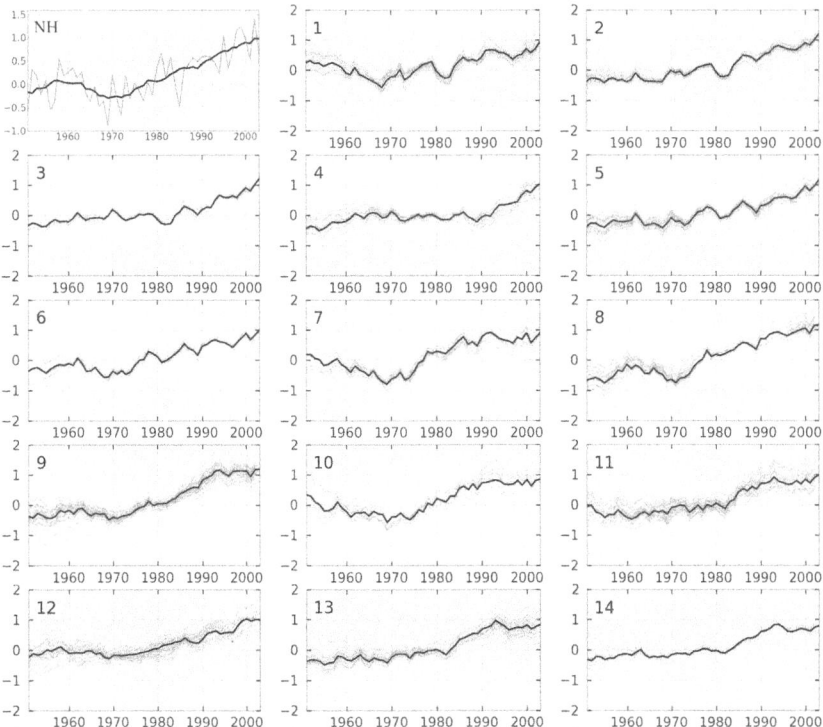

Figure 7.1. | *Temperature Variations at Individual Stations (thin grey lines) and Regional-Mean MAAT (solid black lines) Smoothed with an 11-year Running Filter*

Table 7.1. | Linear Trend Coefficient for Regional-Mean Air Temperature and Precipitation for the 1976–2010 Period

Region	Temperature, °C/100 years					Precipitation, mm/month/100 years				
	Winter	Spring	Summer	Fall	Annual	Winter	Spring	Summer	Fall	Annual
1	9.2	2.9	4.3	5.0	5.3	17	19	8	5	14
2	11.4	3.7	4.9	4.7	6.3	10	15	4	−8	7
3	9.2	5.1	6.5	6.8	6.9	−20	−2	−41	8	−13
4	4.4	3.2	5.7	6.0	4.9	−3	20	−18	42	11
5	6.9	4.5	3.9	7.3	5.8	4	31	−23	−20	−1
6	5.7	4.0	1.8	7.1	4.5	5	40	−3	−6	9
7	4.8	4.7	2.2	3.5	3.4	9	24	2	3	11
8	3.2	7.8	0.9	3.3	3.5	12	16	10	1	10
9	5.0	7.2	5.2	2.4	4.7	7	8	17	16	11
10	5.0	7.4	2.7	2.8	4.7	1	5	17	9	7
11	2.5	5.4	5.3	6.4	5.1	−2	12	23	24	14
12	−0.9	8.9	4.1	8.5	5.4	−9	18	−21	5	1
13	4.5	4.2	2.3	5.1	4.1	8	7	−8	2	3
14	5.9	2.9	2.8	4.6	4.0	11	45	34	−36	15
Average	5.8	5.0	3.8	5.3	5.0	4	17	−2	3	6

Note: Highlighted lines designate regions in the Russian North (see Map 7.1 for region key)

mary of results illustrating seasonal temperature and precipitation trends. All trends were calculated using the 1976–2010 data.

Data in Table 7.1 indicate year-round warming trends all over Russia except for Chukotka (region 12), where winter temperatures have exhibited large interannual variations with near-zero or slight negative trends in the past three decades. Rates of change and seasonal features are not uniform over regions. Winter warming is most pronounced in North-European Russia (region 1), with the temperature trend up to 9.2°C/100 years, nearly twice the trends in summer and fall (4.3 and 5.0°C/100 years, respectively), and more than three times that of spring (2.9°C/100 years). The maximum in seasonal trends shifts gradually from winter to spring along the West–East transect in the Russian Arctic (from the upper to the lower lines in Table 7.1). In Western and Central Siberia (regions 7 and 10), the summer temperature changes are least pronounced, while the rest of the Russian Arctic demonstrates noticeable summer warming.

Air temperature extremes have exhibited much greater changes: annual minimum and maximum temperatures and the temperature range, as well as number of days per year with temperatures above or below certain thresholds (Meleshko 2008). Over large territories in the Russian Arctic, minima rose at a higher rate than maxima, except for in Chukotka (region 12), where annual minima decreased in accord with colder winters. The largest annual temperature minima trends over the past three decades were detected in North-European Russia (14–26°C/100 years) and in Central Siberia (10–14°C/100 years). At the same time, annual temperature maxima did not change in Siberia. Elsewhere in the Russian Arctic, trends never exceeded 10°C/100 years and on average were about 6°C/100 years. Apparently, one can say that rather than getting warmer, the regional climate in Russia is getting less cold.

Annual amounts of precipitation have increased everywhere in the Russian Arctic, with large regional and seasonal variations. Similar to the air temperature pattern, seasonal precipitation maxima have shifted from the cold to the warm period along the West–East transect, with a pronounced snowfall increase in North-European Russia and West Siberia and mostly summer precipitation increase in east Siberia (Table 7.1). In the context of our study, snowfall is particularly important, due to its profound impacts on the urban environment.

There is compelling observational evidence that during the twentieth century, warming patterns in the Russian Arctic, like the rest of the Arctic, have been more pronounced than in other regions of the world. Warming has accelerated in the past thirty years, with the

MAAT northward of 60°N rising at approximately twice the global rate, a phenomenon known as "Arctic amplification" (AMAP 2011). One of the consequences of the concentrated regional warming is a dramatic decline of sea ice in the Arctic Ocean, with an average shrinkage rate of 13.4 percent per decade in the period 1978–2015. In September 2012, Arctic sea ice decreased to a record low level since satellite observations began in 1979, reaching 3.6 million km^2 (about half of the average of the 1980s and 1990s). Remarkably, the period from 2007 to 2011 has had the second through fifth most pronounced warming on record. Sea ice is getting thinner and younger; about 70 percent of it is one to two years old, and 95 percent is younger than five years (AMAP 2011). These trends are projected to increase in the future, opening new opportunities for navigation along the Northern Sea Route (NSR), but as Scott Stephenson discusses in Chapter 8, these accessibility increases will also bring new challenges to the transit regimes of the Russian Far North.

Regional Climatic Projections for Russia

It is possible to develop shorter-term climate projections by extrapolating the trends displayed in Figure 7.1 and the data in Table 7.1 over the next few years. These data characterize statistics of the current trends, which are prone to changes with time, and thus are not necessarily illustrative of the longer-term climate variations at decadal and centennial scales. Long-term projections must include a greater range of data, which can be accomplished by using general circulation models (GCMs). This section provides an overview of recent results from the CMIP5 family of comprehensive GCMs that were used in the preparation of the Fifth IPCC report. The experimental design of the CMIP5 was presented by Tailor, Stouffer, and Meehl (2012). This section evaluates the model's accuracy in a regional context by contrasting results with observations in Northern Russia.

While GCMs are generally acknowledged as the most effective tools for predicting future climate, they should always be viewed critically. Even the most sophisticated computer algorithms are not capable of accounting for the full range of complexity and uncertainty in the climate system. Climate change is governed by the interplay of numerous factors, many of which are stochastic and thus can only be addressed probabilistically, while the mathematical formalism of GCMs is intrinsically deterministic. It is thus not feasible to discern the accuracy of the models through direct comparison with observations of specific months and years sequentially over decadal and centennial time scales. In contrast to weather models, GCMs should not necessarily be judged by their ability to replicate real-time changes in climatic parameters. Instead, one has to look at the difference between statistics in the time series of climatic parameters that are based on observations and model simulations. Robust statistics could only be obtained if continuous observations over a period of twenty-five to thirty years or longer were available, which sets up a minimum time scale at which model results could be viewed as "projections." It could not reasonably be expected that GCM-based climatic projections would be credible at shorter time scales, implying that, at best, they could be used to characterize general future trends. Standardized time periods that have been used in numerous studies are centered on 2030, 2050, and 2080. Alternatively, slices could be bound to a certain period in the future, when the prescribed magnitude of global warming is likely to be reached. The later approach could give insight into the 2°C warmer world (with respect to the preindustrial level), which corresponds to the threshold set by the European Union as a target for its adaptation and mitigation strategies (EU Climate Change Expert Group 2008).

Global climate models have a typical horizontal resolution of about 2° by latitude and longitude, which corresponds to a spatial unit with a size of 200–250 km. Complicating the problem, the results of any individual model for any single grid node are not robust, and the entire pattern contains many unreliable small-scale details, often interpreted as if it is affected by stochastic "noise" (Räisänen and Ylhäisi 2011). This imposes limitations for projecting climatic changes at specific locations, such as individual cities. Similar to observations at individual stations, "noise" may be reduced by averaging several neighboring grids (spatial smoothing), or by applying the same procedure to several models. The ensemble approach is used to minimize the uncertainty of the climate projections. While early studies postulated decreasing uncertainty with the increase in the number of models in the ensemble, more recent papers suggest eliminating outliers, GCMs that demonstrate poor performance in comparison with observations. Model discrimination and construction of optimal ensemble projections for regional studies could be based on the consistency with observations of the specific climatic parameters and indices governing key regional impacts. There is no universal metric for model skills, and numerous procedures have been developed and used to evaluate and rank the models in both the global and regional context. This study uses an ensemble climatic projection that has been optimized for the Russian northern regions using the method described below (Anisimov and Kokorev 2013).

This study used the full set of 36 CMIP5 climate models[3] and evaluated each model's accuracy by comparing calculated trends in the climatic characteristics with observations in the fourteen Russian regions. Tests have been performed using the 1976–2005 data for the seasonal and annual temperatures and sums of precipitation, temperature sums above and below prescribed thresholds, and the dryness index (the ratio of the positive temperature sum to the annual amount of precipitation). Original data have been harmonized by subtracting the "baseline" values averaged over the 1961–1990 period individually for each model. This procedure eliminates systematic biases, which individual models are prone to. Results were averaged over the grid nodes that fall over each of the regions, and compared with the regional observations. Ultimately, models were ranked according to their capability.

Table 7.2 illustrates the disparity between the modeled MAAT trends and observations in the 1976–2005 period for selected regions in the Russian North and in the areas underlain by permafrost. Although not shown in the table, similar results were obtained for other climatic parameters and indexes. Models are classified by their relative errors, defined as the ratio of the difference between the calculated and observed trends of any given climatic parameter to their sum. The threshold for the relative error is set at 0.25 to distinguish between the highly accurate models and those that poorly represent observed regional trends.

Table 7.2. | *Differences Between the Modeled MAAT Trends and Observations in the 1976–2005 Period for Selected Regions in the Russian North, °C/100 years*

Model	Pfrost	1	7	9	10	11	12	13
ACCESS1.3	−1.5	−2.1	0.4	−3.4	−1.7	0.1	0.8	−1.8
ACCESS1-0	0.4	2.6	2.4	−1.7	0.1	0.0	2.5	−1.7
bcc-csm1-1	4.5	7.3	8.0	−2.1	3.9	1.5	−0.6	−0.7
bcc-csm1-1-m	−0.5	−2.4	−0.5	−2.9	0.2	2.8	5.0	2.7
BNU-ESM	0.4	−4.0	0.1	1.0	1.1	2.3	4.4	2.1
CanCM4	−0.2	2.8	2.3	−5.2	−1.2	−0.8	−0.2	−2.1

CanESM2	2.1	0.7	2.3	−1.3	1.1	1.0	3.4	1.0
CESM1−CAM5	−4.4	−5.1	−3.6	−2.8	−3.7	−2.3	−3.9	−3.9
CESM1−FASTCHEM	1.1	0.0	4.4	−2.9	3.8	1.9	−1.2	−1.6
CMCC−CESM	−2.7	−1.0	1.2	−4.7	−0.5	−2.8	−3.9	−5.8
CMCC−CM	1.4	2.3	4.1	4.0	3.3	1.0	−3.3	0.5
CMCC−CMS	−3.6	−5.6	−2.9	−1.2	−3.1	−2.3	−2.0	−1.8
CNRM−CM5	4.4	5.0	7.7	1.4	6.7	4.1	6.7	2.0
CSIRO−Mk3−6−0	0.2	−0.6	2.6	−1.4	1.6	−0.7	−0.8	0.2
EC−EARTH	1.6	−1.7	3.3	−0.9	3.2	0.7	0.6	2.1
FIO−ESM	−0.6	−3.6	2.8	−1.8	2.5	−0.1	−0.2	−2.4
GFDL−CM3	−3.2	−6.5	−1.2	−3.2	−2.9	−3.2	4.0	−1.7
GFDL−ESM2G	1.7	1.6	4.8	−1.4	3.3	3.6	7.4	1.5
GFDL−ESM2M	−1.0	−5.8	−0.6	−0.7	0.6	−0.3	−1.8	−0.8
GISS−E2−H	0.3	3.1	2.5	−4.2	0.1	−0.2	−1.3	0.0
GISS−E2−R	0.3	−0.7	2.2	−1.7	2.5	1.7	0.8	−0.7
HadCM3	1.3	3.0	2.3	−3.0	−0.2	4.1	2.6	0.9
HadGEM2−AO	2.0	1.3	5.3	1.4	5.8	3.4	1.1	0.6
HadGEM2−CC	1.5	−0.3	3.6	−3.0	1.1	1.0	2.3	0.7
HadGEM2−ES	3.0	1.8	2.8	0.4	2.9	3.9	3.9	2.3
inmcm4	−3.3	−5.6	−1.1	−5.3	−2.6	−3.9	−4.0	−4.6
IPSL−CM5A−LR	−1.0	−1.6	1.0	−0.1	−0.6	0.0	0.1	0.9
MIROC4h	−2.7	−3.5	−1.0	−1.8	−1.1	−0.5	−0.2	−2.8
MIROC−ESM	−2.4	−3.3	−0.4	−1.4	−0.6	−0.9	0.2	−0.6
MIROC−ESM−CHEM	0.4	0.5	2.8	−4.2	0.2	−1.9	−1.0	−2.6
MPI−ESM−LR	−3.8	−8.0	−2.7	−4.1	−3.4	−0.9	0.0	−0.5
MPI−ESM−MR	−0.3	−0.3	1.8	−2.1	0.2	0.9	2.5	−2.5
MPI−ESM−P	−1.1	−2.0	2.7	0.4	2.6	0.1	−1.3	−2.3
MRI−CGCM3	−4.6	−5.7	−3.6	−5.9	−3.6	−3.0	−1.4	−4.2
NorESM1−M	0.8	2.4	3.0	0.3	0.7	−0.3	3.8	0.0
NorESM1−ME	−2.9	0.7	1.5	−5.3	−3.8	−6.8	−3.9	−5.4

Source: Based on data from CMIP5 historical runs

Some questions remain open, such as how to treat models that demonstrate a high accuracy with respect to certain climatic parameters in one region but perform poorly when other parameters in other regions are considered. These instances are displayed as grey cells in Table 7.2, which indicate that the relative model error is above the prescribed threshold for at least one of all tested parameters in the corresponding region. One solution to this problem is to eliminate such models and to combine the remaining ones into the optimal regional ensemble. The other option would be to keep them in the ensemble while assigning them a smaller weight. In the latter case, efforts should be made to avoid biased weighting. As was demonstrated by Weigel et al. (2010), if weights do not appropriately represent the full range of skills with respect to various parameters as well as associated uncertainties in these parameters, ensembles with weighted models perform on average worse than those in which all models have equal weights. When actual and modeled climate variability is large, as is the case in our study regions, more information may be lost by inappropriate weighting than potentially could be gained if weighting is optimal. Largely due to these considerations, this study uses only ensembles with equally weighted models (Weigel et al. 2010). Another question is how to evaluate the projected changes of climatic parameters for individual cities. In accord with what has already been said, the ideal would be to calculate climatic norms using local observational data, and to superimpose the trend averaged over one of the selected large regions.

Plots in Figure 7.2 show the variety of regional-mean MAAT projections from individual CMIP5 models (light grey curves), the ensemble of all 36 models (dark grey curve), and the optimal ensemble of models with the best regional accuracy (black curve) for selected regions in the Russian North, including the areas underlain by permafrost. Hereafter, predictive CMIP5 model runs have been used for the twenty-first century under the high greenhouse gas emission scenarios likely to result from the developing world economy (RCP-8.5) (Riahi, Gruebler, and Nakicenovic 2007). By 2025, the MAAT in most regions of the Russian North is projected to rise by 2–3°C relative to the 1961–1990 norm, and by approximately 4°C through the mid-twenty-first century. Interestingly, except for Western Siberia (region 7), the optimal ensemble predicts higher rates of warming than the average over all models. Although the differences between the ensemble-means are small, the optimal ensemble has an added value in narrowing the range of uncertainty in climate projections by eliminating those based on GCMs with poor regional accuracy.

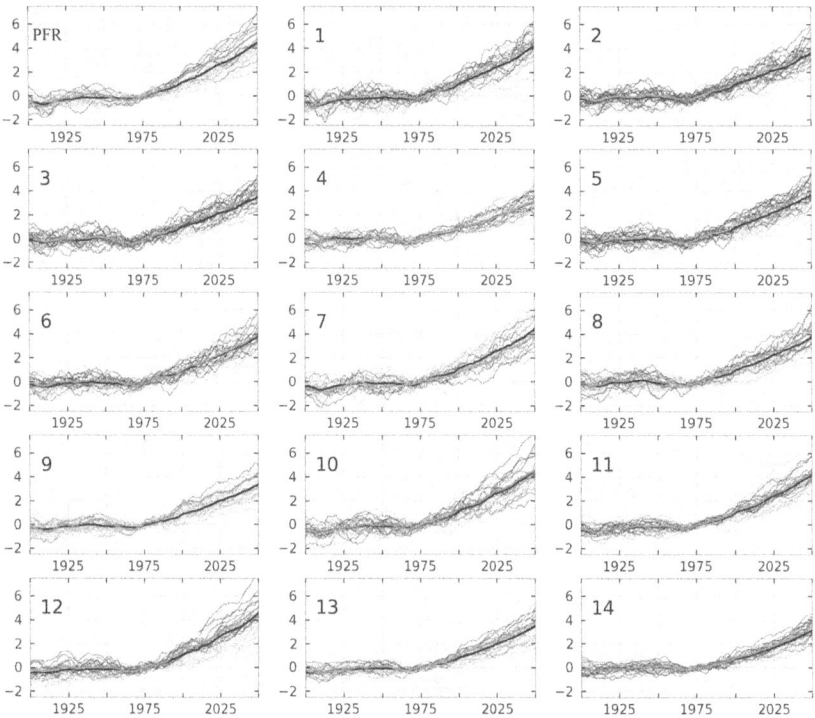

Figure 7.2. | *Regional-Mean MAAT Projections from Individual CMIP5 Models (light grey curves), Ensemble of all 36 Models (dark grey curve), and Optimal Ensemble of Models with the Best Regional Skills (black curve)*

Climate Change and Urban Sustainability in the Russian Arctic

Urban development in the Russian Arctic has been largely driven by the exploration of natural resources, as well as by the need to support marine and river transport operations and maintain defense systems in the coastal zone and northern seas. Of the approximately 370 villages and settlements in the Russian high Arctic (tundra zone), more than 80 percent are located in the coastal zone or in close proximity to large rivers (Anisimov 2010). Unlike other northern countries, the Russian Arctic is characterized by intensive urban developments, and has cities with more than 100,000 population, most of whom are permanently employed and serve the needs of regional industries (Table 7.3).

Table 7.3. | *Russian North City Characteristics*

Region	Cities	Population	Key local industries
North-European Russia (reg.1)	Arkhangelsk	355,800	timber production, river port
	Kholmogory	4,592	timber production
	Severodvinsk	190,083	ship repair
	Shenkursk	5,548	timber production and manufacturing
	Vorkuta	70,548	coal
	Syktyvkar	253,432	forestry, timber
	Ukhta	99,600	coal
	Severomorsk	50,060	ship repair, fisheries
	Apatity	59,672	aluminium
	Zapolarny	15,825	iron
	Kandalaksha	35,654	shipyard
	Kirovsk	28,625	apatite
	Kovdor	18,820	iron
	Monchegorsk	45,361	nickel
	Murmansk	304,508	sea port serving the NSR
	Nikel	12,756	nickel
	Olenegorsk	23,072	nickel
	Naryan-Mar	42,844	river and sea port
	Petrozavodsk	265,263	machinery production, forestry
West Siberia (reg.7)	Nefteyugansk	138,000	oil
	Nizhnevartovsk	262,600	oil
	Nyagan	57,101	oil, forestry
	Surgut	322,900	oil and gas
	Khanty-Mansiysk	85,029	regional administrative center
	Tyumen	609,650	oil and gas
	Nadym	47,360	oil and gas
	Novy Urengoy	112,192	oil and gas
	Noyabrsk	107,210	oil and gas
	Salekhard	46,552	regional administrative center
	Urengoy	10,070	gas

Southern Siberia (reg. 9)	Irkutsk	600,000	energy, coal and lignite, aircraft, heavy engineering
Central Siberia (reg.10)	Norilsk	176,189	nickel, copper, cobalt, non-ferrous metals
	Dudinka	23,923	sea port, part of NSR
Sakha-Yakutia (reg.11)	Lensk	24,373	river and road transportation, port
	Neryungri	62,333	coal
	Yakutsk	267,983	administrative center
	Mirniy	35,994	diamonds
	Aldan	23,371	gold
Chukotka (reg.12)	Anadyr	13,529	gold, coal, non-ferrous metals
	Magadan	95,925	gold, silver, non-ferrous metals
Far East (reg. 13)	Blagoveshchensk	219,861	machinery production, forestry

Source: Anisimov 2010, updated

To the extent that the public focuses on the problem, its perception of climate change's impacts on the urban environment in the Arctic typically focuses on potentially detrimental consequences for the infrastructure built upon thawing permafrost. Numerous examples of climate- and permafrost-related infrastructure failure have been presented in academic and popular publications and are discussed in detail by Dmitry Streletskiy and Nikolay Shiklomanov's Chapter 9 in this volume. Meanwhile, climate impacts are much broader than just those associated with thawing permafrost, and include both challenges and opportunities. Besides thawing permafrost, serious concerns are associated with changes in the freshwater ice and the hydrological regime.

Human settlements in the Russian Arctic, including cities with a population of over 50,000, are often located in close proximity to rivers, which serve as essential transportation routes linking them with other parts of the country. A particular concern with riparian settlements is the risk of floods caused by ice jams, which occur in all Arctic and sub-Arctic regions. Jams develop abruptly, lead to much higher water levels than freshets caused by thermal-driven snow melt, and may have potentially catastrophic consequences. Although not discussed here, floods have many positive ecological impacts, such as the replenishment of riparian ecosystems in the flood plain with water and nutrients (AMAP 2011). The mechanism of ice jam floods is well understood, and besides numerous scientific publications, has been detailed in the SWIPA assessment report (AMAP 2011). The uneven onset of ice break-up in spring leads to ice aggregation at certain locations followed by elevated water levels and flooding in the upstream segment of the river. The subsequent release of ice jams is a related concern, associated with a steep water wave characterized by high flow velocity and significant destructive potential (Beltaos and Burrell 2008). This phenomenon is exemplified by the catastrophic flood of the city Lensk on the Lena River in May 2001, which is detailed in the case study further in this section.

Despite its risks, not all impacts of climate change in the Arctic are negative. A less severe climate will reduce the demand for heating energy, and current observations and model-based projections suggest that hydropower generation would benefit from reduced ice periods and increased runoff in winter, when energy demand is at its annual maximum (AMAP 2011). The duration of the ice period on rivers in the circumpolar North has been decreasing since the 1970s on average by 12 days/100 years with up to 4 times greater rates in the high Arctic. Statistically, an increase in autumn and/or spring air temperature of 2 to 3°C leads to a 10- to 15-day shift in freeze-up and/or break-up of the river ice in the Arctic (AMAP 2011). Lengthening of the ice-free period on rivers and in the northern seas opens new opportunities for transportation over water. In the period 1980–1999 the entire NSR was open for navigation up to 45 days per year. With the current dramatic decline of sea ice extents in the Arctic Ocean, navigation has become more feasible. According to model projections, by the mid-twenty-first century, there will be up to three months per year suitable for navigation along the NSR. (See Scott Stephenson's Chapter 8 in this volume for a detailed discussion of transportation in the Arctic.)

The potential benefits for water transportation are in part balanced by the reduced usability of ice roads currently serving the supply needs of remote settlements in the Arctic, which would otherwise remain isolated in winter. In the coming decades many of the ice roads and river crossings may become economically unfeasible necessitating significant investments into the development of all-weather roads (AMAP 2011) (see Chapter 9). Climate change will also have some positive implications for the health of Arctic residents, such as a decrease of injuries, cold-related diseases, and mortality associated with extreme cold temperatures (ACIA 2005).

Snow

In thinking about the implications for urban sustainability, the role of snow is threefold. First, it has direct implications for city maintenance costs given the substantial resources required for regular snow cleaning of highways, streets, airport runways, and roofs. Snow affects the performance of the engineered systems of buildings in permafrost areas by blocking the ventilation spaces designed to prevent northern buildings from warming (and thus deforming) underlying permafrost. Buildings are heated throughout the winter, and special care is needed to minimize the warming effect on permafrost and keep the temperature of the frozen ground below the threshold incorporated in the design of the foundation, achieved through constructing open basements surrounded by fences with ventilation windows, and by means of passive ground coolers. Obligatory maintenance operations in northern city management include regular snow removal around the buildings to allow the circulation of cold air in the basement through the ventilation windows.

Second, snow acts as a thermal insulator, and as such is an important factor governing the ground thermal regime and the state of permafrost. Ground temperature under snow cover is several degrees higher than it would be if the surface were exposed to the atmosphere. Lastly, the amount of snow accumulated over the cold season and the spring temperatures are two main factors governing the severity (peak water level rise) and duration of annual freshets, as opposed to floods due to ice jams.

In summation, a longer snow period and deeper snow cover are associated with increased operational expenses in the urban environment, leading to the warming of permafrost through higher thermal insulation during the cold period, enhanced risks for infrastructure on pile foundations, and increased peak water levels during spring freshets. Decreasing snow depth and duration are thus favorable for urban Arctic regions.

Shmakin (2010) analyzed observed changes in regional snow characteristics and calculated the sums of solid precipitation over the cold period, which is defined in his paper as that with daily average air temperatures below 1°C. In the Arctic, this period lasts from midfall to late spring. Results indicate up to a 30 percent increase in the cumulative amount of snowfall in Chukotka between the 1951–1980 and 1989–2006 timeframes, mostly due to spring snowfall, which outweighed a decline in winter precipitation over the same time period. In West Siberia and in the northeast of the European part of Russia, cumulative snowfall increased by 10 percent–20 percent, whereas in eastern Siberia, snowfall decreased up to 20 percent (Shmakin 2010).

This study uses the same method described by Shmakin (2010), and applies it to CMIP5 model results to construct projections of snow period length and cumulative amount of snowfall (in water equivalent) over a period with daily average temperatures below 1°C. This study also uses projections of the cumulative precipitation sums for the fixed cold period from October to May to characterize the intensity of seasonal precipitation. Results for each of the study regions in the Russian North are presented in Figure 7.3. Snow accumulation depends on the seasonal precipitation intensity and duration of the snow period. The total amounts of precipitation in the period from October to May (Figure 7.3A) are projected to increase with a rate varying from 0.82 mm/y in Central Siberia (region 10) to 1.62 mm/y in North-European Russia (region 1). The proportion of liquid precipitation (rain) in this period will progressively increase with time due to warming. Depending on the interplay between the increasing total precipitation and fraction of it falling as rain, maximum snow depth is projected to decrease in some regions (12 and 13), and increase in the others (Figure 7.3B). The duration of the snow period (not shown here) is projected to shorten everywhere at a rate varying from −0.70 d/y and −0.61 d/y in North-European Russia and West Siberia (regions 1 and 7), to less than −0.43 d/y in Yakutia and Southern Siberia (regions 11 and 9).

Changes in Air Temperature

There are many ways by which air temperature affects cities in the Russian Arctic. Cumulative degree-days of thawing (ddT—the total number of days with temperatures above 0°C) govern the state of permafrost and its ability to support structures built upon it. The temperature regime directly affects the budget expenditures on heating over the winter period, measured as the duration of the heating period (dH), which is defined as the period when daily air temperature is below 8°C, and heating degree-days (Hdd), defined as the cumulative sum of the daily differences between the physiological comfort temperature (prescribed at 18.3°C) and ambient air temperatures over the heating period. Such temperature limits are set in Russian Federal regulations for heating standards (Construction code, 2003), and they generally align with the principles of heating regulation used in other countries (e.g. Day et al. 2003; Isaac and Vuuren 2009).

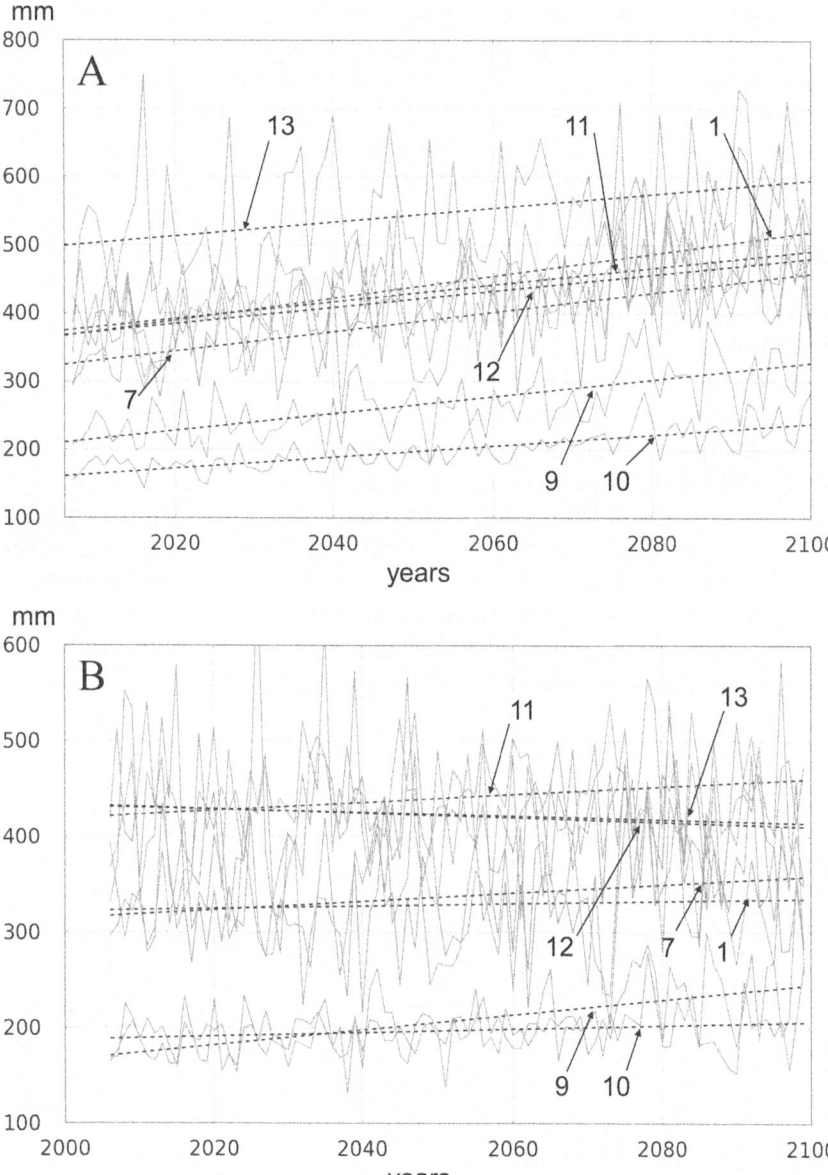

Figure 7.3. | *Projected Regional Changes in the Cumulative Amounts of Precipitation in the Period October–May (A), and in the Snowfall Period with Temperatures Below 1°C (B)*

Note: Calculations are based on CMIP5 ensemble climate projection under the RCP-8.5 emission scenario.

Projected changes in the degree-days of thawing are presented in Figure 7.4. According to these data, the expected rate of ddT rise (trend) decreases from west to east in the Russian Arctic, varying between 16.5 degree-days per year (°C d/y) in North-European Russia (region 1) and 15.7°C d/y in West Siberia (region 7), to 11–12°C d/y in the Central Siberia, Yakutia, and Chukotka (regions 10 through 12 in Figure 7.4). By 2050, the Russian Arctic will be accumulating much more heat during the summer. Except for Chukotka (region 12) all regions will be characterized by ddT rates higher than those of present-day North European Russia (region 1), where permafrost is relatively warm. By the end of the century the regional-mean ddT everywhere in the Russian North is projected to rise well above the current ddT level in the warmest of all permafrost regions.

Khlebnikova, Sall, and Shkolnik (2012) used observational data and calculated changes in the demand for heating energy between the two periods of 2001–2010 and 1981–2000. According to results of this study, Hdd dropped by 200–300°C d (about 5–8 percent) in North-European Russia, by less than 200°C d (0–2 percent) in West Siberia, and by 200–500°C d (2–5 percent) elsewhere in the Russian North. Projected changes in the heating regime have been addressed by several Russian publications, some of which have been summa-

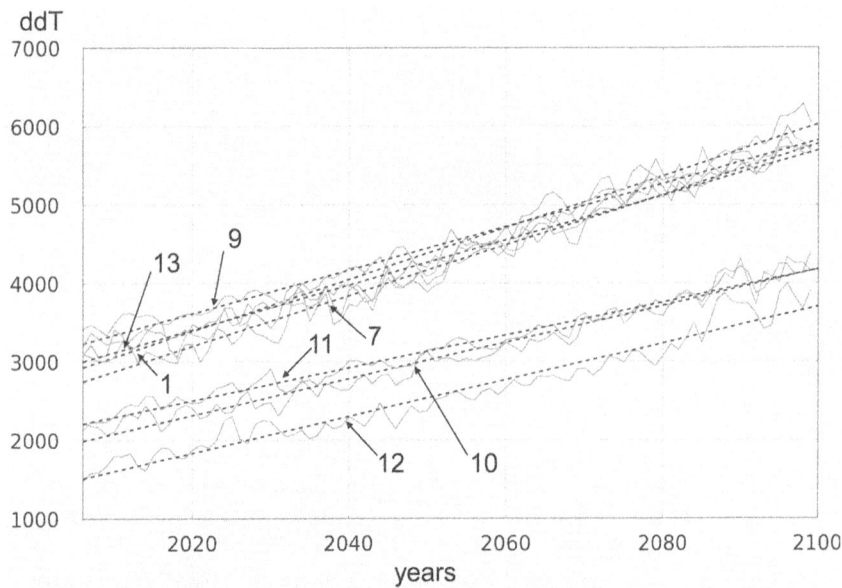

Figure 7.4. | *Projected Changes of Thawing Degree-Days (ddT), °C d*
Note: Calculations are based on CMIP5 ensemble climate projections under the RCP-8.5 emission scenario.

rized in English by Anisimov and Vaughan (2007). According to these results, demand for heating energy in the Russian Arctic is expected to decrease up to 15 percent under projected mid-twenty-first century climate conditions, whereas the duration of the period when heating in the northern cities is needed will decline by up to one month.

These findings have been updated with the most recent CMIP5 climate projections, and scaled down to selected regions in the Russian North. Projected characteristics of the heating regime for specific regions are presented in Figure 7.5. The rate at which the demand for heating energy is decreasing in the Russian Arctic (Hdd trend, Figure 7.5B) varies from −21.8°C d/y in North-European Russia (region 1) to −27°C d/y in West Siberia (region 7) and −30.2 to 31.7 in Yakutia and Central Siberia (regions 11 and 10) due to the cumulative effect of less severe winters and the shortening of the heating period (dH, Figure 7.5A).

River Floods

Climatic warming will lead to changes in the frequency, duration, and severity (peak water levels) of floods on northern rivers. The current situation, available data, and trends in the frequency and severity of ice jams on northern Russian rivers have been analyzed in papers by Buzin (2007) and Buzin and Kopaliani (2007, 2008). A summary of their findings in English is given in the SWIPA assessment report (AMAP 2011).

In 2007 Buzin and Kopaliani predicted in the near-term period of 2010–2015 that ice-jam floods on major Siberian rivers would affect larger proportions of the total channel length, become more frequent, and have much higher peak water levels than in the baseline period before 1977 (Map 7.2). One of the driving factors governing such changes is the delay of ice break-up at latitudes 58° to 60°N due to distinct temperature gradients that often occur in this zone in spring. Particular concerns were associated with the Severnaya Dvina, Sukhona, Vug, and Pechora Rivers that traverse urbanized areas in north-European Russia. Many cities located along these rivers, including Shenkursk, Kholmogory, Arkhangelsk, Naryan-Mar, and Veliky Ustug, were likely to be affected by an increased frequency and severity of ice-jam floods, although projected near-term changes were smaller than in Yakutia and Chukotka. The frequency of floods in North-European Russia was projected to increase by 20 percent at most, while peak water levels were expected to increase by 35–50 percent. Our analysis suggests reasons why these explanations were not correct.

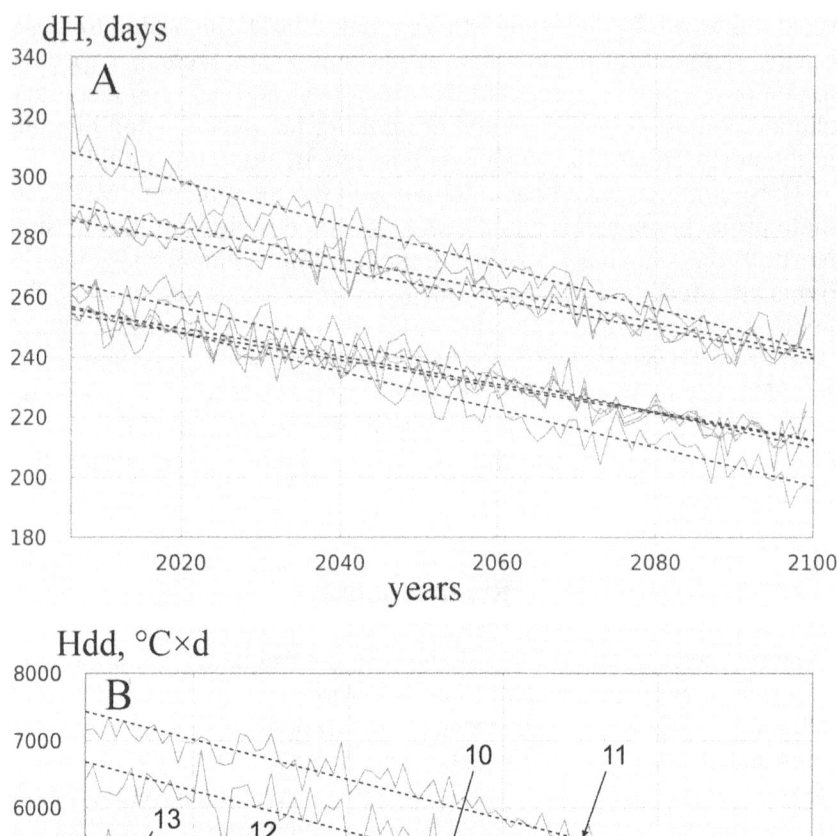

Figure 7.5. | *Projected Regional-Mean Changes in the Characteristics of the Heating Regime*

Note: A—heating period length. B—heating degree-days. Calculations are based on CMIP5 ensemble climate projection under the RCP-8.5 emission scenario.

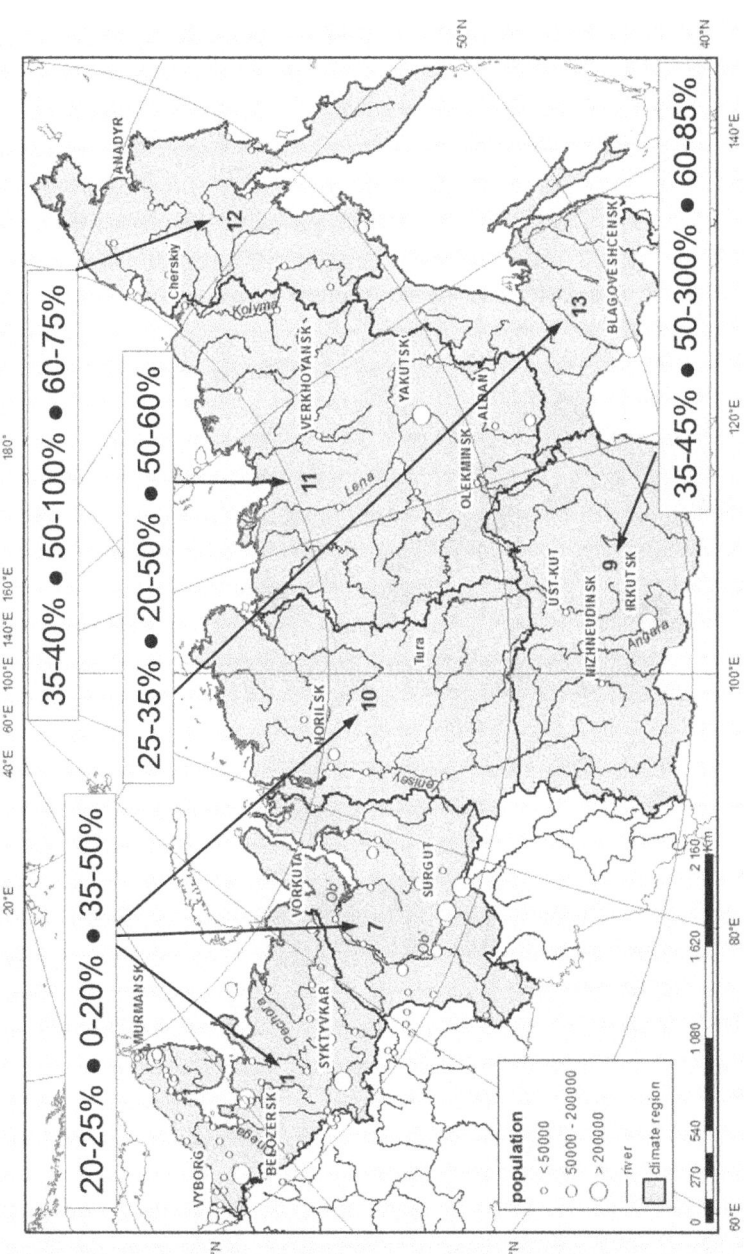

Map 7.2. Projected 2010–2015 Changes in the Characteristics of Ice-Jam Floods Relative to the Baseline Period 1946–1977

Note: Numbers indicate projected relative changes in the following parameters: flooded fraction of the channel length, frequency of flooding, and maximum water level.
Source: Modified from Buzin (2007) and Buzin and Kopaliani (2007, 2008)

Ice dynamics are largely governed by the thermal gradients along rivers (AMAP, 2011). For northward-flowing rivers, such as those in the Russian Arctic, the onset of warm temperatures and ice break-up come earlier upstream, ultimately leading to ice jams downstream. Climatic warming could change the situation to the better by lowering the thermal gradient along rivers in spring, if the rate of warming in the downstream segments exceeds those upstream. Alternatively, if climatic warming leads to the enhancement of the thermal gradients along rivers, there is a potential for the increased frequency and severity of ice jam floods.

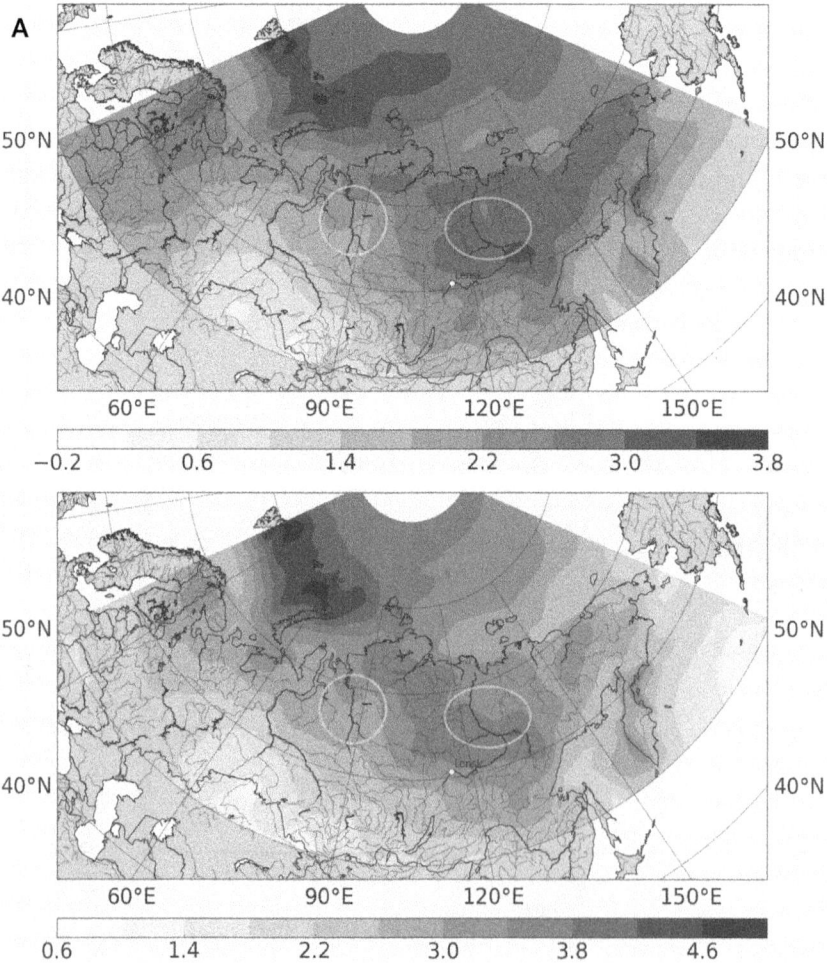

Figure 7.6. | *Projected Air Temperature Changes for the Spring Break-up Period (May) Relative to the 1961–1990 Norm*
Note: Figure A—2006–2015, Figure B—2016–2025

Projected spring temperature changes are an important spatial feature, especially those in May, which is the most common month for ice-jam floods to occur in the Russian North. The maps in Figure 7.6 illustrate the projected May temperature changes relative to the 1961–1990 norm. Results for several decadal time samples (only two of which are shown in Figure 7.6) consistently show a tendency toward greater warming in the downstream channel segments on all northern rivers. This will lead to a reduction in the current temperature gradient so that ice breakup will be increasingly driven by thermal factors rather than mechanical ice action. Such changes are likely

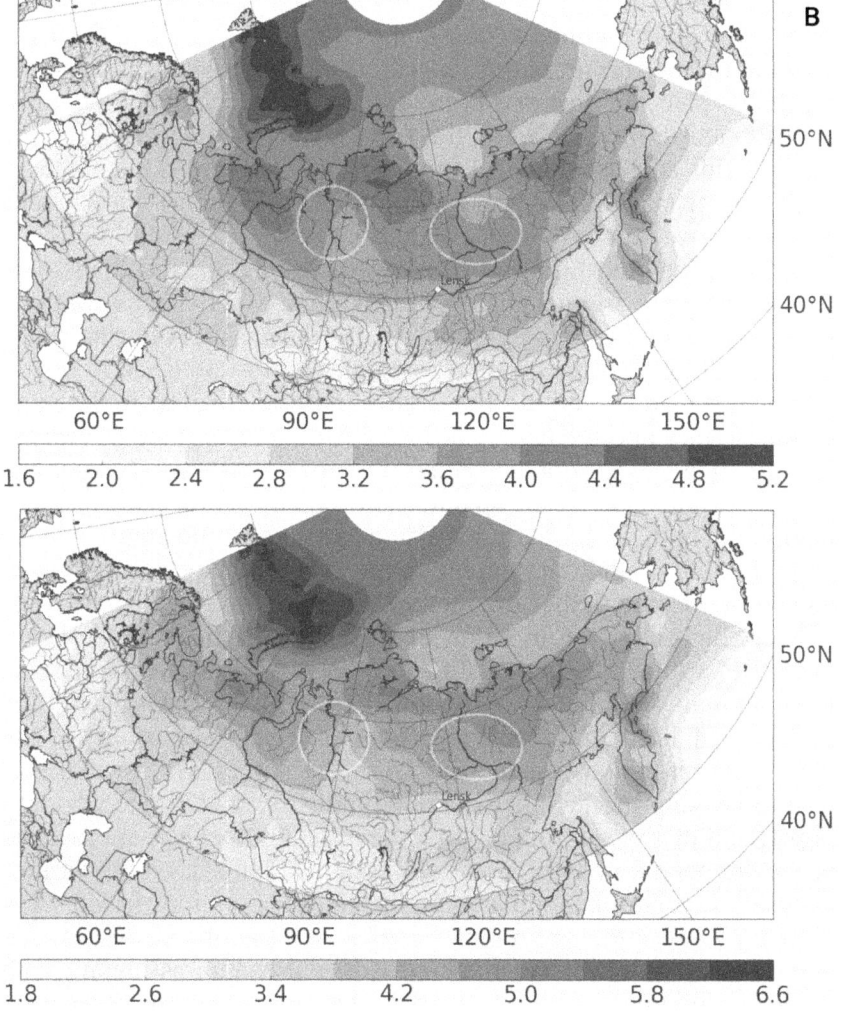

to reduce the probability of ice jams nearly everywhere in the Russian North, except for a few locations indicated by circles on the maps, including the segment of the Lena River in the vicinity of Yakutsk and a small segment of the Yenisei River. This result does not support the conclusion of Buzin (2007) and Buzin and Kopaliani (2007, 2008) regarding the increased frequency of ice-jam floods on larger proportions of river channels in the next several years.

Case Study: 2001 Flood in Lensk

The history of Lensk[4] is illustrative of man's failure to plan for natural risks in the construction of cities in the North. Lensk existed for many years in harmony with nature; however, planning decisions made in the 1950s placed a new part of the city on a flood plain. In 2001, this decision led to disaster when the Lena River flooded and destroyed much of this construction. Unfortunately, the authorities did not learn from this mistake: after a visit from President Vladimir Putin, the city was quickly rebuilt in the same location.

Existing records show that Lensk was founded no later than 1663. The settlement's original name was Mukhtuia, which means "Big Water" in the Evenk language. In the 1730s, during Vitus Bering's second expedition to Kamchatka, a post office was established in Mukhtuia on the road linking the cities of Irkutsk and Yakutsk. In the nineteenth century and the early part of the twentieth century, the village received political prisoners exiled from European Russia. The population of no more than 500 people worked at the post office and in maintaining the road, in agriculture (raising cattle and cultivating potatoes and vegetables), and trapping fur. In the 1930s, the local economy expanded to include timber production and inland river transportation.

Everything changed in 1956, when the village became the headquarters for the construction of the new city of Mirnyi, set up to mine the diamonds found in a volcanic kimberlite pipe 231 km inland (by air) from Mukhtuia. The village benefitted from its proximity to the river Lena and soon became a base in the supply system that was vital to develop the lucrative diamond industry. Within one year, the population increased from 2,000 to nearly 8,000 people, and peaked at 30,900 people in 1992. Most of the population served the various needs of the diamond industry and its mining operations in Mirnyi.

In 1963 Lensk met the requirements to achieve official designation as a city. By that time it had a well-developed urban infrastructure, with multistory buildings, river links, an airport, and a 280-km all-weather road connecting it to the recently built diamond-producing center of Mirnyi.[5] Proceeds from the diamond industry financed the development of the city, but the builders put little thought into the kind of urban planning required to protect the new structures from natural disasters, particularly floods. Due to this inappropriate planning, in 1956 construction of the new part of the city took place on a flood plain.

In May 1998 the abrupt onset of warm weather caused rapid snowmelt and led to severe flooding of the Lena River. On 17 May, the water level peaked at 17 meters above the norm, damaging many structures. The estimated economic loss in Lensk and smaller towns on the banks of the Lena totaled 872.5 million rubles (equivalent to approximately US$148 million at the 1998 exchange rate).

The next year, the river flooded again, nearly reaching the height of the flood in 1998. It should have been possible to learn lessons from the experiences of 1998 and later years and, if it were not possible to protect the city from the rising waters, to at least minimize any unavoidable losses. Yet to prevent flooding in 2001, the Yakutian authorities considered it sufficient to spend only 15 million rubles (approximately US$530,000), or 0.05 percent of the republic's budget. Unfortunately, the flooding in May 2001 turned out to be much worse than in previous years. As a result, the federal government had to pay 6 billion rubles (US$214 million) to repair the damage caused by the waters and rebuild the almost completely destroyed city. Given the extent of the damage, the republic's government was unable to pay for the losses, so the federal government was required to spend 400 times what had been spent earlier on preventive measures.

Already by March 2001, the Lena basin showed all of the major risk factors for future flooding. The previous winter had been cold and snowy, with snow packs greatly exceeding the norm. The thickness of the ice on the Lena was on average 20–30 cm greater than usual, and in places exceeded the multiyear average by 1.5 to 2 times, reaching 3 meters. In the beginning of May the difference in temperature in the flow of the Lena in Irkutsk Oblast and Sakha (Yakutia) Republic reached 20°C, making possible the accumulation of a large quantity of water in the upper reaches of the Lena River and its tributaries and an earlier ice melt in this area (with a normal river flow, the ice melt would start later). The water began to flow with unusual speed: in the course of a day, the ice "jumped" 250 kilometers and tumbled onto Batami Island, located 40 km downstream from Lensk, where an 80km ice jam formed.

The water began to rise at the speed of 35–40 cm/hour. On 17 May, it reached a critical point, and the dike protecting Lensk gave way, allowing water to flow directly onto the city's streets. By noon on 18 May, the level of the water in the city had risen by more than 20 meters. All of the ships that had been in the river were crushed by the ice and sunk. At 4:55 pm, the ice logjam was broken and ice began to flow on all parts of the Lena River. As a result, the level of water began to drop with a speed of 1.5–2.5 cm/minute. In Lensk and the surrounding area (Batamai, Saldykel', Nyuya, Natora, and Turukta), 3,331 homes were completely destroyed and 1,831 required extensive reconstruction. In addition, 396 km of electrical lines, 164 substation transformers, 470 km of communication wires, and 5 radio transmission stations were damaged. It was necessary to restore 184 km of roads, 2 bridges, 7 healthcare centers (clinics, hospitals, natal care centers), and 26 schools and child care facilities. Approximately 31,000 people lived in the affected area; of these, 8 people died and more than 20,000 suffered property damage. The furnaces were destroyed in the houses that were flooded (with a cost of tens of thousands of rubles each) and often all that was left were bits of clay. The Lena River turned into a carpet of floating firewood heading for the Laptev Sea. During winter, every Yakutian resident needs 30–40 cubic meters of wood for heating; with the loss of the old wood to the flood, the locals had to gather new wood to replace it. The residents' ice-cellars also flooded, ruining the food supply of frozen fish kept there. After the flood, it was typical to find huge chunks of ice on the streets of the city. Many small homes were destroyed or completely torn from their foundations.

Could this disaster have been avoided? Many scientists suggest that one problem was that the authorities did not carry out the annual dredging that had been conducted in previous years. Before regular dredging began along the Lena, the water rose above the critical level near the city of Yakutsk once every 7–12 years (1917, 1924, 1933, 1946, 1958, and 1966). But when the dredging was regularly performed (1970–1990) and the river was at least 3 meters deep near Yakutsk, rising waters were not a problem. During those years, just in the part of the river lying below Yakutsk, dredgers removed 5–5.5 million cubic meters of soil just to clear one navigation route. With the end of the dredging work and the gradual transition to the natural state of the river, there began to be a series of spring floods (1998, 1999, and 2001). Consequently, it can be concluded that human inaction, in addition to the natural tendencies of the river, was a cause of the 2001 flood in Lensk.

The subsequent development of events was even more interesting. On 24 May, Russian President Vladimir Putin arrived in Lensk and held a meeting with officials to discuss how to deal with the consequences of the flood, stating at the session opening, "First of all, I want to know what was and was not done to prevent the tragedy, and if something was not done, why not?" (Gafutulin 2001). The Russian leader described the situation in the affected area as difficult and announced that 30,000 had suffered as a result of the flood, adding that the destroyed parts of Lensk would be rebuilt in the same place. On that day, the State Duma adopted a special ruling about Yakutia. Ultimately, Lensk was rebuilt by the deadline established by the Russian president and government—1 October. The decision of the president and government to restore the city on its previous location, apparently adopted to confirm the thesis that Russians are stronger than nature, continues to arouse amazement and questions from specialists, especially given the possibility that the dramatic events of 2001 could be repeated.

Conclusion

The results presented in this chapter illustrate that cities in the Russian North will be facing numerous challenges and opportunities in association with climate change. Data in Table 7.4 summarize the projections of the regional-mean climatic and hydrological characteristics that have been selected for analysis in this chapter due to their potential impacts on the urban environment. While these data draw a general pattern of the rates of regional changes, there are large

Table 7.4. | *Projected Changes in the Regional-Mean Climate Characteristics*

Region	ddT, °C d/10y	Hdd, °C d/10y	Snow period, days/10y	flood frequency, %	water level, %
North-European Russia (reg. 1)	164.9	−218.3	−6.4	0–20	35–50
West Siberia (reg.7)	157.4	−270.9	−5.8	0–20	35–50
Southern Siberia (reg. 9)	131.9	−209.5	−4.5	50–300	60–85
Central Siberia (reg. 10)	119.9	−316.6	−5.3	0–20	35–50
Sakha-Yakutia (reg. 11)	110.5	−302.8	−4.9	20–50	50–60
Chukotka (reg. 12)	118.2	−311.1	−7.0	50–100	60–75
Russian Far East (reg. 13)	135.6	−235.8	−4.6	20–50	50–60

gradients of baseline climatic conditions within each region, and the particular impacts on specific cities depend on local conditions.

Caution should be used, as due to the presence of the south-north gradient within each region, these regional-mean data are not suitable for direct application to specific locations such as individual cities. To get insight into smaller levels of spatial details, one can use city-specific norms calculated through observational data from local weather stations and overlay it with regional trends, such as those in Figure 7.4. The latter are likely to be representative for the entire region since homogeneity with respect to the rate of climatic change has been one of the key considerations behind the regionalization developed in this study (Map 7.1), applicable to nearly all results presented in this section, except for projections of floods, which are specific to river basins.

Overall, this chapter shows that Russia's north can expect extensive changes in its climate and that urban planners will have to take these changes into account. While the changes will differ from place to place, it is clear that there will be, on average, warmer temperatures. Subsequent chapters will examine the consequences of these climate changes for transportation and urban infrastructure.

Oleg Anisimov is Professor of Physical Geography at the State Hydrological Institute in St. Petersburg.

Vasily Kokorev is a researcher at the State Hydrological Institute, Climate Department, in St. Petersburg.

Notes

1. This research was supported by the Russian Science Foundation, project 14-17-00037.
2. Available at http://www.iea.ru/article/kioto_order/15.12.2009.pdf.
3. Available at http://pcmdi9.llnl.gov.
4. http://www.gorodlensk.ru.
5. http://www.mojgorod.ru/r_saha/lensk/.

References

ACIA. 2005. *Arctic Climate Impact Assessment.* Cambridge: Cambridge University Press.
AMAP. 2011. "Arctic Climate Issues 2011: Changes in Arctic Snow, Water, Ice and Permafrost." SWIPA 2011 Overview Report. Oslo: Arctic Monitoring and Assessment Program.

Anisimov, O.A. (ed.). 2010. *Environmental and Socio-economical Impacts of Climate Change in Permafrost Regions: Predictive Assessment Based on Synthesis of Observations and Modeling.* Moscow: Greenpeace.
Anisimov, O.A. and V.A. Kokorev. 2013. "Constructing an Optimal Climate Ensemble for Evaluation of the Climate Change Impacts on the Cryosphere," *Ice and Snow,* no. 1: 83–92 (in Russian).
Anisimov, O., V. Kokorev, and Y. Zhil'tsova. 2013. "Temporal and Spatial Patterns of Modern Climatic Warming: Case Study of Northern Eurasia," *Climatic Change.* doi:10.1007/s10584-013-0697-4.
Anisimov, O. and D. Vaughan. 2007. "Polar Regions," in *Climate Change 2007: Impacts, Adaptation, and Vulnerability. Contribution of Working Group II to the Fourth Assessment Report of the Intergovernmental Panel on Climate Change.* Cambridge: Cambridge University Press, pp. 653–86.
Beltaos, S. and B. Burrell. 2008. "Climatic Aspects," in S. Beltaos (ed.), *River Ice Breakup.* Highlands Ranch, CO: Water Resources Publications, pp. 377–406.
Budyko, M.I. 1962. "Polar Ice and Climate." *Proceedings of the Academy of Sciences of the USSR,* Geographical series no. 6: 3–10 (in Russian).
———. 1969. *Climatic Change.* Leningrad: Hydrometeoizdat (in Russian).
Buzin, V.A. 2007. "Risk of Floods on the Rivers of Russia: The Analysis of Tendencies and Possible Development of a Situation in the Near Future." *Sb. Dokladov mezhdunarodnoi nauchnoi shkoly* [Collection of reports of the international scientific school] "Modelling and Analysis of Safety and Risk in the Complicated Systems," St. Petersburg, pp. 496–501 (in Russian).
Buzin, V.A. and Z.D. Kopaliani. 2007. "The Inundations on the Rivers of Russia with the Current Weather-Change Trends. The Up-to-Date Trends to the Alteration of Climate," *Uchenye zapiski RSHMU,* no. 5: 43–54 (in Russian).
———. 2008. "Ice Jam Flooding on the Rivers of Russia: Risks of Their Occurrence and Forecasting." In *Sb. dokladov koferentsii upravleniya vodnoresursnymi sistemami vekstremal'nykh usloviyakh* [Collection of reports at the conference on control over water—resource systems under extreme conditions], Moscow, 4–5 June, pp. 282–83 (in Russian).
Construction code 23–02–2003. "Heat Conservation in Buildings." Moscow: Gosstroy of Russia (in Russian).
Day, A.R., I. Knight, D. Gunn, and R. Gaddas. 2003. "Improved Methods for Evaluating Base Temperature for Use in Building Energy Performance Lines," *Building Services Engineering Research and Technology* 24(4): 221–28.
EU Climate Change Expert Group "EG Science." 2008. *The 2°C Target. Information Reference Document.* Retrieved 3 July 2016 from www.climateemergen cyinstitute.com/uploads/EU_2C_2008.pdf.
Gafutulin, N. 2001. "Lensk vozroditsya na prezhnem meste," *Krasnaya zvezda,* 25 May. Retrieved 16 July 2014 from http://old.redstar.ru/2001/05/25_05/1_01 .html.
Hansen, J.R. Ruedy, M. Sato, and K. Lo. 2010. "Global Surface Temperature Change," *Reviews of Geophysics* 48, RG4004, doi:10.1029/2010RG000345:1–29.
Isaac, M. and Detlef P. van Vuuren. 2009. "Modeling Global Residential Sector Energy Demand for Heating and Air Conditioning in the Context of Climate Change," *Energy Policy* 37: 507–21.

Khlebnikova, Y.I., I.A. Sall, and I.M. Shkolnik. 2012. "Impacts of Regional Climatic Changes on Infrastructure," *Meteorology and Hydrology*, no. 12: 19–35 (in Russian).

Khon, V., I. Mokhov, M. Latif, V. Semenov, and W. Park. 2010: "Perspectives of Northern Sea Route and Northwest Passage in the Twenty-First Century," *Climatic Change* 100(3): 757–68.

Meleshko, V.P. (ed.). 2008. "Assessment Report on Climate Change and Its Consequences in the Russian Federation." Vol. 1. Moscow.

Räisänen, J. and J.S. Ylhäisi. 2011. "How Much Should Climate Model Output Be Smoothed in Space?" *Journal of Climate* 24: 867–80.

Riahi, K., A. Gruebler, and N. Nakicenovic. 2007. "Scenarios of Long-Term Socioeconomic and Environmental Development Under Climate Stabilization," *Technological Forecasting and Social Change* 74(7): 887–935.

Shmakin, A.B. 2010. "Climatic Characteristics of Snow in Northern Eurasia and Their Changes in the Past Decades," *Ice and Snow*, no. 1: 43–58 (in Russian).

Simpson, R.D. 2009. "The Economics of Resources and the Economics of Climate," *Journal of Natural Resources Policy Research* 1(1): 103–6.

Solomon, S., D. Qin, M. Manning, Z. Chen, M. Marquis, K.B. Averyt, M. Tignor, and H.L. Miller. (eds.). 2007. "Climate Change 2007: The Physical Science Basis. Contribution of Working Group I to the Fourth Assessment Report of the Intergovernmental Panel on Climate Change." Cambridge: Cambridge University Press.

Tailor, K.E., R.J. Stouffer, and G.A. Meehl. 2012. "An Overview of CMIP5 and the Experiment Design," *Bulletin of the American Meteorological Society* 93: 485–98.

WCIOM. 2008. "Grozit li nam globalnoe poteplenie?" WCIOM Press Release no. 1049, 17 September. Retrieved 16 July 2014 from http://wciom.ru/index.php?id=459&uid=10708.

———. 2010. "Kto potushil pozhary: dozhd, bozhya pomoshch ili MChS?" WCIOM Press Release no. 1580, 13 September. Retrieved 16 July 2014 from http://wciom.ru/index.php?id=459&uid=13818.

Weigel, Andreas P., Reto Knutti, Mark A. Liniger, and Christof Appenzeller. 2010. "Risks of Model Weighting in Multimodel Climate Projections," *Journal of Climate*: 4175–91. DOI: 10.1175/2010JCLI3594.1

CHAPTER EIGHT

Access to Arctic Urban Areas in Flux
Opportunities and Uncertainties in Transport and Development

Scott R. Stephenson

Introduction

The IPCC-projected 1.1–4.8°C of global surface warming over the next century is expected to precipitate unprecedented environmental change (IPCC 2013).[1] This global temperature increase is likely to be strongly amplified in the Arctic as melting sea ice and reduced snow cover allow solar radiation to be increasingly absorbed by polar seas and landmasses (ACIA 2004a). As a result, the Arctic is often viewed as a bellwether for global climate change, as thawing permafrost, melting glaciers and ice sheets, and sea ice recession signal the beginning of a warmer climate regime in the northern high latitudes.

These physical changes have powerful implications for human systems. Scholars from a variety of fields have extensively documented the social impacts of climate change in areas such as water security, agriculture, fisheries, and transportation (Parry et al. 2007). Of these, transportation is of particular salience in the Arctic, as both geophysical and infrastructural factors have altered northern accessibility in complex and often competing ways. For example, thawing permafrost and poor design have contributed to degradation of built structures (Streletskiy, Shiklomanov, and Nelson 2012a, 2012b) while receding sea ice and the proliferation of ice-classed ships have afforded new navigation potential in the Arctic coastal seas, particularly along the Northern Sea Route (NSR).

Accessibility represents a central concept in a geographic understanding of the world. It reflects an organization of space facilitating social, economic, and political activity through networked systems of interaction, highlighting the affordances and limitations of the physical landscape on which these activities take place. Accessibility describes simultaneously the function and relative importance of nodes of human activity, the linkages that connect them, and the flows that give them meaning (Taaffe and Gauthier 1973). Furthermore, places are drawn differentially into existing economic and social networks according to their specific geography of accessibility (Agnew 2002). Physical, economic, and political affordances (or constraints), such as topography, transportation infrastructure, and border crossings, are central to understanding the "relative opportunity of interaction and contact" among human communities (Johnston 1986).

Economic development in the Arctic depends critically on access to remote resources and communities. Because resources are necessarily tied to specific places, development requires reliable and cost-effective connections between extraction sites and consumer markets. These connections are mediated by a dynamic, rapidly changing physical environment and sparse, environmentally sensitive infrastructural networks. Understanding accessibility in this region requires attention to two climatically sensitive transportation systems: marine transport in ice and terrestrial winter roads. Changes to these systems have important implications for community resupply (inward transport) and resource exploitation (outward transport), leading some Arctic settlements to become increasingly integrated with global economic networks, and others less integrated and effectively more remote.

This chapter will examine the economic implications of recent and projected future changes in Arctic accessibility in response to a warmer climate regime. While marine navigation seasons are expected to lengthen with continued sea ice recession, the degree to which increased access will enable economic and urban development in the region is highly uncertain. Furthermore, climate change is likely to decrease terrestrial access in winter owing to reduced winter road potential. First, this chapter will review the science underlying projections of growing Arctic marine accessibility and the implications for resource extraction and shipping. The chapter will then discuss the physical and economic challenges impeding Russian Arctic oil and gas development and transit shipping, followed by an examination of the impact of reduced terrestrial winter access on northern industries and communities. The chapter will close with a discussion of future development scenarios in Arctic Russia and infrastructural requirements for the creation of a shipping network based on the NSR. This last section shows that national development priorities, rather than existing climate conditions, are the main driver for investments in the Arctic and its related urban development, though the state of the sea ice and rivers will have a major impact on the focus of the investment.

Resource and Shipping Potential with Reduced Sea Ice

There has been a growing consensus over the past decade that reduced sea ice extent will increase maritime access throughout the Arctic (ACIA 2004a; Arctic Council 2009). Satellite data collected since 1979 reveal a robust downward trend in summer ice extent and thickness (Comiso et al. 2008; Kwok et al. 2009; Maslanik et al. 2007; Serreze, Holland, and Stroeve 2007; Stroeve et al. 2008, 2012b) with record lows in 2007 and 2012 (NSIDC 2012a). Reductions have also occurred in winter, with the March 2012 maximum extent falling 614,000 km^2 below the 1979–2000 mean (NSIDC 2012b). The loss of older, thicker multiyear ice coverage has been especially widespread, declining ~15 percent per decade since the 1980s (Comiso 2012; Maslanik et al. 2011; Polyakov, Walsh, and Kwok 2012). While climate models exhibit considerable variability in timing and magnitude of ice loss, all models project a negative trend in ice extent throughout the twenty-first century (Christensen et al. 2007; Stroeve et al. 2012a; Vavrus et al. 2012; Wang and Overland 2009; Zhang and Walsh 2006).

Despite its adverse consequences for the global environment, reduced sea ice is often portrayed as promoting economic opportunity and integration for the region. Projections of a first ice-free Arctic Ocean in the coming decades (Wang and Overland 2009) have fueled interest in trans-Arctic maritime routes linking major world markets. Numerous studies have highlighted the considerable potential distance savings of Arctic routes compared with traditional routes through the Suez and Panama Canals (Arbo et al. 2012). While the NSR has been utilized commercially for decades, recent research suggests that new theoretical "trans-polar" routes representing the shortest marine distance between Europe and East Asia may become technically viable by mid-century (Smith and Stephenson 2013). Circumnavigating overland canals and other narrow waterways such as the straits of Malacca and Hormuz bestows the added advantage of avoiding strategic "choke points" that may be blocked for political reasons or unreliable due to pirate activity (EIA 2011).

Oil and gas are currently produced in four Arctic states (Canada, Norway, Russia, and the United States), which together account for approximately 28 percent of world oil and 46 percent of gas output (BP 2011). Large deposits were discovered at Tazovskoye field in Russia in 1962 and in Prudhoe Bay in Alaska in 1967 (Østreng 2012). Since then, 61 large oil and gas fields have been discovered in territory north of the Arctic Circle in these four countries. Of these, 42 are located in Russia, 11 in Canada, 6 in Alaska, and 1 in Norway (Budzik 2009). These fields combined with estimated undiscovered deposits represent a significant proportion of world total potential reserves. A widely cited USGS study estimates that the Arctic contains 13 percent of the world's undiscovered oil and 30 percent of its gas, approximately 84 percent of which is under exclusive state control less than 200 nautical miles from shore (Gautier et al. 2009). One-third of the undiscovered oil (30 billion barrels) is in Alaska, while one-third of the undiscovered gas is in Russia's West Siberian Basin. While Greenland is not currently a petroleum exporter, its offshore East and West Basin provinces are estimated to exceed 16 billion barrels of oil (Gautier et al. 2009). In aggregate, Arctic hydrocarbons represent one of the most significant remaining unexploited sources of nonrenewable energy. Recognizing the lure of such potential, a recent report by insurance giant Lloyd's of London (2012) declared that "the Arctic is likely to attract substantial investment over the coming decade, potentially reaching $100 billion or more."

Despite the relative low in current oil prices (below $50/barrel as of June 2016), there are reasons to believe that Arctic supplies, however remote and difficult to access, may again become economically viable. In July 2008, the price of oil reached a historic high of $147/barrel, many times higher than the assumed "natural price" of $22 to $28 four years earlier (Yergin 2012). Following a sharp decrease coinciding with the global recession, prices resumed their upward trajectory largely due to fundamental increases in demand from OECD countries and rapidly industrializing economies in China and India. World energy demand is projected to increase 36 percent from 2008 to 2035, with China alone accounting for 75 percent of the increase (International Energy Agency 2010). Owing to their relatively cheap cost of implementation, fossil fuels are expected to cover approximately 80 percent of world demand by 2030. These projections suggest that regardless of petroleum commodity financialization, the world oil market has undergone a structural shift toward prices that reflect higher aggregate demand (Yergin 2012). With production at some unconventional US shale oil plays already approaching maturity, new exploration projects in the Arctic may become increasingly attractive.

In response to new economic prospects, sea traffic in the Arctic has increased in recent years to levels unprecedented in the post–Cold War era. The landmark Arctic Marine Shipping Assessment reported approximately 6,000 ships operating in the Arctic in 2004 (Arctic Council 2009). About half of the voyages were Great Circle Route transits between Asia and North America south of the Bering Strait, while the rest were predominantly destinational, driven by resource exploitation, community resupply, and tourism. These drivers are expected to increase considerably over the next decade; by 2020, it is projected that annual demand for resupply operations in Canada alone will exceed the capacity of the current fleet (Arctic Council 2009). The NSR, in particular, has seen a number of recent milestones. In August 2011, a Panamax-class tanker carrying 61,000 tons of gas condensate sailed the NSR in a record eight days, followed by the first-ever transit by a Suezmax-class supertanker seven days later (Pettersen 2011a, 2011b). In 2013, 71 ships transited the route carrying 1.35 million tons of cargo, a recent record (NSR Information Office 2013). Voyages are occurring increasingly later in the navigation season, as demonstrated by the first-ever transit by a liquid natural gas (LNG) carrier in November 2012 (McGrath 2012).

Arctic states are well aware of the jurisdictional implications of an increasingly accessible northern ocean. Should the Arctic coastal states (Canada, Denmark/Greenland, Norway, Russia, United States) submit successful claims to extend the limits of their exclusive economic zone under UNCLOS Article 76, their marine boundaries would encompass even greater areas of newly accessible ocean currently classified as international waters. In 2001, Russia claimed an area of nearly 1.2 million square kilometers representing the greatest theoretically claimable area of the Arctic states. The UN Commission on the Limits of the Continental Shelf did not accept or reject the claim, instead recommending additional research before a decision could be made. Russia resubmitted its claim in 2015. If granted, Russia would economically control an area of sea floor running from the Siberian shore to the North Pole. While Russia would not have the power to enforce nondiscriminatory environmental regulations granted under UNCLOS Article 234 in the newly claimed area, it would have sovereign control of any hydrocarbon deposits beneath the seabed.

As evidence of increasing maritime accessibility in the Arctic has grown, studies modeling the present and future potential of Arctic navigation have proliferated (Verny and Grigentin 2009; Khon et al. 2010; Liu and Kronbak 2010; Smith and Stephenson 2013; Stephenson et al., 2013; Stephenson, Brigham, and Smith 2014; Stephenson and Smith, 2015). Such studies often describe future navigation potential in terms of scenarios of ice and/or economic conditions. For example, Khon et al. (2010) found significant differences among climate models in their ability to simulate navigation season length in the NSR and NWP. Other studies (Corbett et al. 2010; Paxian et al. 2010; Peters et al. 2011) projected rising greenhouse gas and particulate emissions from future Arctic marine traffic, highlighting the potential of shipping as a positive feedback on climate warming.

Framing their analysis in geographical terms, Stephenson, Smith, and Agnew (2011), Stephenson et al. (2013), and Stephenson, Brigham, and Smith (2014) examine regional changes in Arctic marine access under a range of climate scenarios. In accordance with future reductions in sea ice concentration and thickness, large areas of the Arctic will become significantly more accessible to Polar Class vessels by midcentury (Table 8.1), including a majority of the Russian coastal seas in summer (Figure 8.1). Navigation seasons in the NSR are projected to exceed three months on average throughout the twenty-first century using such vessels (Table 8.2). Consistent with observed sea ice seasonality, accessibility is greatest in summer (July–October), though increases relative to early-century conditions occur year-round (Table 8.1; Figure 8.1). This suggests that in addition to expected future increases in summer traffic, the Arctic may experience limited vessel activity in relatively ice-free coastal seas during winter. Winter access will remain limited to ice-strengthened vessels, however, as even the most aggressive climate models predict winter ice cover throughout the twenty-first century (Stroeve et al. 2012a). As the Arctic transitions to a seasonally ice-free state, the permanent loss of perennial multiyear ice will increase year-round accessibility for ships capable of operating in first-year ice.

Tempering Optimism about Arctic Exploitation

While a seasonally ice-free Arctic appears likely in the coming decades, whether attendant marine access will enable economic and urban development in the region is highly uncertain. Russian political and industry leaders often promote an optimistic view of Arctic development facilitated by the NSR, in which trans-Arctic shipping heralds a new era of direct integration with global logistics industries

Table 8.1. | *Annually Averaged Changes in Marine and Terrestrial Accessibility by Midcentury (2045–2059) vs. Baseline (2000–2014)*

	Winter road-accessible land area (km²) (2000–2014)[a]	Winter road-accessible land area (km²) (2045–2059)[a]	Change in winter road accessible land area (km²)[a]	Maritime-accessible ocean area (km²) (2000–2014) Current EEZ[b]	Maritime-accessible ocean area (km²) (2045–2059) Current EEZ[b]	Change in maritime-accessible ocean area (km²) Current EEZ[b]	Maritime-accessible ocean area (km²) 2000–2014 Extended EEZ claims[b]	Maritime-accessible ocean area (km²) (2045–2059) Extended EEZ claims[b]	Change in maritime-accessible ocean area (km²) Extended EEZ claims[b]
Canada	3036089	2636279	−399810 (−13%)	3783485	4497573	+714088 (19%)	3795341	4996567	+1201226 (32%)[c]
Finland	37204	21969	−15235 (−41%)	84389	84389	0 (0%)	N/A	N/A	N/A
Greenland	212919	189519	−23400 (−11%)	1550086	1982799	+432713 (28%)	1571997	2151113	+579117 (37%)[c]
Iceland	2488	450	−2038 (−82%)	754275	755825	+1550 (< 1%)	N/A	N/A	N/A
Norway	20102	9929	−10173 (−51%)	1997040	2037728	+40688 (2%)	2326232	2374703	+48471 (2%)[c]
Russia	4817544	4199588	−617956 (−13%)	6073261	7062516	+989255 (16%)	6246559	8065798	+1819239 (29%)[d]
Sweden	31433	16944	−14490 (−46%)	159300	159300	0 (0%)	N/A	N/A	N/A
USA (Alaska)	449294	321108	−128185 (−29%)	3475270	3660049	+184779 (5%)	3518895	3893787	+374892 (11%)[d]
High seas	N/A	N/A	N/A	451211	2282190	+1830980 (406%)	108838	296848	+188010 (173%)[e]
Total	8607073	7395786	−1211287 (−14%)	18328316	22522369	+4194053 (23%)	17567861	21778815	+4210955 (24%)

Notes: [a]2000-kg GVWR vehicle; [b]Canadian Type A vessel; [c]theoretical claim (International Boundaries Research Unit 2008); [d]pending claim; [e]unclaimed

Source: Adapted from Stephenson, Smith, and Agnew 2011.

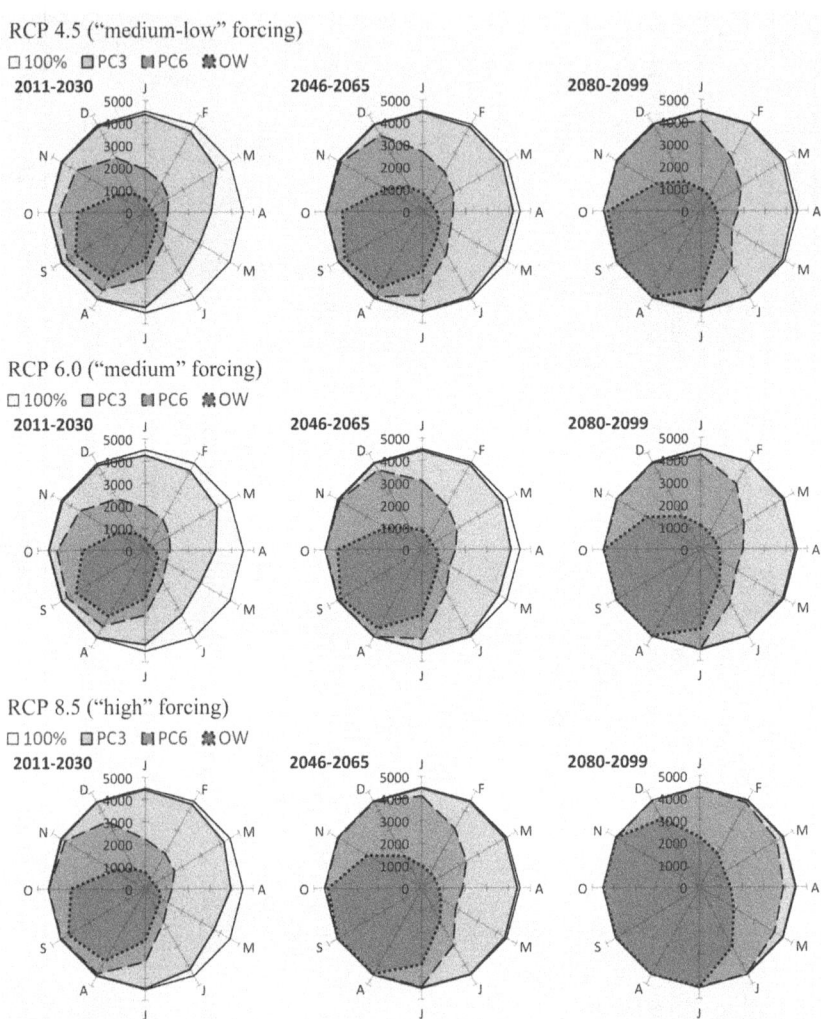

Figure 8.1. | *Total Ship-Accessible Marine Area in the Russian Maritime Arctic (1000 km²) as Driven by Climate Forcing Scenario (RCP 4.5/6.0/8.5), Time-Averaging Window (2011–2030, 2046–2065, 2080–2099), and Vessel Class (Polar Class 3 [PC3], Polar Class 6 [PC6], Open-Water [OW])*

Note: Outer circles signify 100 percent year-round access.
Source: Stephenson et al. 2013.

Table 8.2. | *Average and Standard Deviation (Italics) of Navigation Season Length in Summer (July–October; Max Season Length: 123 days) by Early (2011–2030), Mid- (2046–2065), and Late Century (2080–2099) to Polar Class 3 (PC3), Polar Class 6 (PC6), and Open Water (OW) Vessels along the Northern Sea Route, under RCP 4.5/6.0/8.5 Climate Forcing*

		PC3		PC6		OW	
RCP 4.5	2011–2030	111	*18*	98	*26*	81	*27*
	2045–2065	120	*6*	113	*13*	101	*21*
	2080–2099	121	*4*	117	*9*	109	*15*
RCP 6.0	2011–2030	110	*18*	99	*24*	85	*26*
	2045–2065	117	*11*	108	*19*	97	*23*
	2080–2099	122	*3*	120	*6*	115	*10*
RCP 8.5	2011–2030	118	*11*	109	*20*	97	*26*
	2045–2065	122	*3*	119	*8*	112	*13*
	2080–2099	123	*1*	122	*1*	121	*3*

Source: Stephenson et al. 2013.

and resource markets. Putin summarized this viewpoint at the 2011 Arctic Forum: "The shortest route between Europe's largest markets and the Asia-Pacific region lie across the Arctic ... I want to stress the importance of the Northern Sea Route as an international transport artery that will rival traditional trade lanes in service fees, security and quality. States and private companies who choose the Arctic trade routes will undoubtedly reap economic advantages" (Bryanski 2011). Similarly, Russia's plan to invest heavily in Arctic oil and gas development is underpinned by an assumption that newly accessible offshore supplies will become commercially viable. Russia has a strong incentive to increase oil and gas supplies as heavy reliance on exports has left its economy and government highly vulnerable to price fluctuations. Russia relies on oil and gas for two-thirds of its exports, 20 percent of its GDP, and 60 percent of its state budget (Hulbert 2012). Shale gas discoveries in North America have driven down gas prices, and oil prices remain below the level required to balance Russia's federal budget ("The Beginning of the End of Putin," *The Economist*, 2 March 2012). Many of Russia's older and larger fields are in long-term decline, and new production is increasingly coming from smaller and more expensive fields (Oil and Energy Trends 2012). New Arctic offshore fields represent a means of maintaining output in the face of such production declines, and an opportunity to supply emerging markets in Asia via the Northern Sea Route (Stephenson and Agnew 2016). According to current Rosneft Chairman Igor Sechin, up to 40 percent of Russian oil output will come from new offshore Arctic and Black Sea prospects by 2030 (Westbrook 2012).

There are reasons to temper such optimism. Arctic oil and gas plays remain among the most expensive to develop in the world and carry considerable environmental risks, particularly offshore (ACIA, 2004b; Mulherin et al. 1996; USARC Permafrost Task Force 2003; Verma, White, and Gautier 2008). Despite projecting substantial future investment in Arctic projects, Lloyd's of London (2012) cautioned that drilling in the Arctic "constitutes a unique and hard-to-manage risk," prompting German bank WestLB to announce that it would not finance offshore projects in the region (Kroh 2012). Data from Alaska's North Slope indicate that the cost of drilling an onshore well in the Arctic may be as much as 640 percent higher than the US average (EIA 2008). Furthermore, costs have risen sharply in the last decade as new exploration has turned toward increasingly remote and marginal fields. From 2000 to 2005, onshore drilling costs in Alaska rose 564 percent, compared with 165 percent for the United States as a whole (American Petroleum Institute 2006). Offshore wells are many times more expensive in the Arctic (~$60 million; Chukchi Sea) than at lower latitudes (~$7 million; Gulf of Mexico) (Østreng 2012). Where infrastructure is underlain by permafrost, climate change–induced subsidence and thickening of the seasonal active layer increase construction and maintenance costs further (Cole, Colonell, and Esch 1999; Hinzman et al. 2005; Larsen et al. 2008; Streletskiy, Shiklomanov, and Nelson 2012a, 2012b).

Climate change presents numerous challenges to navigation and coastal development even as it reduces overall ice extent. Warmer temperatures and higher humidity promote fog formation and reduce visibility, especially in an expanded navigation season in late fall when daylight periods are short. Reduced overall ice concentration can cause thick multi-year ice in the central Arctic Ocean to drift toward coastal shipping lanes (Howell and Yackel 2004; Melling 2002). Multiyear ice is harder than first-year ice and can cause significant hull damage in all but the most ice-strengthened ships. A longer ice-free season increases wind fetch in open water (Barber et al. 2010), accelerating ice drift and enhancing the risk of ice collisions. Larger ice-free areas also strengthen passing storm systems through increased heat transfer to the atmosphere, leading to larger and more frequent storm surges along the coast (Vermaire et al. 2013). In addition, permafrost thaw and reduced landfast ice accelerate coastline erosion, creating construction and maintenance challenges for ports and other land-based infrastructure.

The environmental impact of an oil spill in the Arctic would be especially severe. Sea ice, low visibility, high winds, rough seas, and cold temperatures complicate all aspects of spill response (Pew Environment Group 2010). Oil may remain on or within sea ice after a spill only to be released months later (Atlas, Horowitz, and Busdosh 1978; Martin 1979). Such "trapped" oil is difficult to track and cannot be reclaimed using conventional methods (AMAP 2007; Pew Environment Group 2010). Furthermore, vulnerability to spills is not limited to areas of extractive activity or oil shipment corridors. Many vessels, particularly those operating in the eastern Arctic, continue to rely on low-cost heavy fuel oil (HFO). Accidents involving these ships may cause spills regardless of the cargo on board (Det Norske Veritas 2011). In a follow-up report on the 2010 Macondo disaster, the US National Oil Spill Commission (2012) noted that while cleanup methods such as in-situ burning and mechanical recovery of oil have been demonstrated in ice, they have not been successfully tested in the extreme weather conditions likely to be present in the Arctic. In addition, the report cited a lack of preparedness within the Coast Guard to deal with a serious drilling accident in the Arctic, and the "substantial controversy" remaining over the adequacy of spill response plans and containment capability. Likewise, a US federal report found gaps in Shell's spill response plan for drilling in the Beaufort Sea, noting its failure to account for the unique risks of offshore development in icy conditions (US Government Accountability Office 2012). The binding Agreement on Cooperation on Marine Oil Pollution Preparedness and Response signed in May 2013 under the auspices of the Arctic Council is an important step toward multilateral environmental stewardship in the region, but is limited to international coordination of spill response efforts rather than prevention and safety measures.

Shell's troubled offshore Alaska program well illustrates the difficulties of drilling in the Arctic by even relatively well-equipped companies. In July 2012, Shell's drilling rig *Noble Discoverer* slipped anchor and nearly ran aground near Dutch Harbor, well south of the Arctic Circle. Drilling was further delayed in August when its oil spill response barge failed to receive Coast Guard certification, and when the containment dome it carried was damaged during testing in the Puget Sound. Forced to abandon deepwater drilling in 2012, Shell settled for drilling two "top holes" in the Beaufort and Chukchi Seas. While being towed back to Seattle in late December, its *Kulluk* rig broke free from its cables and ran aground in the Gulf of Alaska. Although no spill occurred, the mishap cast doubt on Shell's safety management procedures, leading an Interior Department review to conclude that Shell had failed in a wide range of basic operational tasks (Broder 2013). Though low oil prices were the primary cause of Shell's exit from the Alaskan Arctic in 2015, uncertainty about ice conditions, technical requirements, and the regulatory regime were important contributing factors.

The fate of oil and gas development in Russia is likewise ambiguous. Offshore exploration may proceed relatively quickly in Russia due to targeted economic policies and fewer environmental regulations, while the Obama Administration has no need to rush Arctic development at a time when domestic onshore shale gas is plentiful. However, these same resources are changing the global supply structure for petroleum to the possible marginalization or exclusion of Arctic resources. New unconventional deposits such as shale gas, shale oil, and oil sands, as well as new conventional production made possible by horizontal drilling techniques, have dramatically increased supplies from relatively accessible and infrastructurally advanced fields around the world. Such plays are meeting demand that otherwise would have justified investment in Arctic projects such as Shtokman, which has been postponed indefinitely due to the US shale gas revolution. While for the moment China appears to be a large and reliable market for west Siberian oil and LNG, China is also looking to expand domestic energy production in order to curb its reliance on imports in the long term. Combined with the high costs and long lead times of Arctic production, these new supplies indicate that Arctic hydrocarbons will be economic only with continued rising demand for oil and gas worldwide, rather than in select markets only. With assumed increases in supply from OPEC and non-OPEC countries, Lindholt and Glomsrød (2012) found Arctic supplies to account for only 8–10 percent of global petroleum production by 2050. While Russia accounts for the vast majority of Arctic gas production in their study, the authors project a lower Arctic share of world production in 2050 than today due to cheap and abundant reserves from the Middle East coming on stream. Studies like these illustrate the dependence of Arctic development on exogenous factors such as global commodity prices and the geography of supply chains.

In light of these challenges, Russian cities hoping to capture a share of revenue from petroleum and transit activities must adjust expectations toward a reality of persistent high operating costs and an uncertain pace of development. While Arctic marine traffic is certain to increase as the navigation season expands, Russian claims that the

NSR will soon rival the Suez Canal as a global shipping corridor are exaggerated. In the near term, Arctic routes will be ill suited to container shipping, which prioritizes reliability and economies of scale over speed and is increasingly driven by just-in-time production models. However, shipping schedules on Arctic routes must account for unpredictable weather and ice conditions, leading to delays and supply chain disruptions. The length and timing of the navigation season is highly variable owing to interannual variability in the seasonality of sea ice retreat and advance (Belchansky et al. 2004; Laxon, Peacock, and Smith 2003; Stephenson et al., 2013; Stephenson, Brigham, and Smith 2014). Bathymetry also presents a challenge to large-volume shipping and economies of scale, as shallow draft in some straits of the NSR prevents passage by ships carrying 50,000 deadweight tonnage or greater (Moe and Jensen 2010). Medium- and large-capacity ships (for example, Panamax and Suezmax) must sail north of the New Siberian Islands where they are more likely to encounter severe and unpredictable ice conditions. In addition, much of the Arctic sea floor remains poorly mapped, limiting route choice and forcing navigators to rely on early British admiralty charts for bathymetric information in some areas (Joling 2012). Transit shipping in the NSR may be more economically viable for bulk commodities such as oil and minerals, which are less sensitive to arrival dates. These transits, combined with destinational traffic in support of resource extraction, are likely to constitute the majority of Arctic shipping activity in the first half of the twenty-first century. As Norwegian Vice-Admiral Haakan Bruun-Hanssen asserted at the 2013 Arctic Frontiers conference, future Arctic shipping will be driven by "profit, safety, and reliability" (Bennett 2013) rather than regional resource appraisals and melting ice.

Winter Roads

Climate change also has powerful implications for terrestrial accessibility in the north. Because of the scarcity and high cost of permanent roads in the Arctic and sub-Arctic, terrestrial access is often made possible by winter/ice roads. Winter roads are seasonal roads constructed over frozen land, rivers, and lakes using compacted snow, applied ice caps, and ice aggregates (Adam 1978). These roads are constructed either by removing snow from frozen land and water bodies and applying water in layers to form smooth ice (Cardinal 2011) or by packing snow to construct a workable driving surface (Adam 1978). Land cover that remains impassible from spring to fall, such as boggy muskeg and peatlands covering vast areas of northern Russia and Canada, becomes accessible in winter in this way.

Winter roads must be built each year, last for only a few months, and are vulnerable to short-duration warm weather events. However, their substantially lower cost (~$5000/km vs. ~$1 million/km for permanent roads that can transport heavy freight) makes them attractive as precursors to permanent roads in industrial operations or long-term solutions in remote communities (Guyer and Keating 2005; Zaitsoff 2011). As a result, winter roads have become a vital part of the resource development sector throughout the Arctic and sub-Arctic. For example, in northern Canada, permanent road use in summer is subject to regulations that limit loads to as little as 30 percent of a typical winter load (Rollheiser 2011). Winter roads are therefore required for transport of heavy machinery that exceeds these limits. Oil from the Novoportovskoye field on the Yamal Peninsula is currently transported 200 km by ice road to Payuta before being taken southward by rail. Gazprom Neft intends to replace the ice road with a railway connection as production increases (Staalesen 2013).

Winter roads also serve as vital supply links for northern communities (Argounova-Low 2012). In Russia, such connections (usually called "ice roads") often take the form of river ice crossings. For example, an ice road crosses the Ob River linking Labytnangi and Salekhard in the Yamal-Nenets Autonomous Okrug (Tananaev 2013). Yakutsk is connected south to the Amur region via the Lena Federal Highway and east to Magadan via the Kolyma Federal Highway by ice roads that cross the Lena River. These highways are open year-round and connect to Yakutsk by ferry in summer and by ice roads in winter. Hundreds of such crossings link communities throughout Siberia, of which thirty to forty are opened officially every winter in Yakutia. The majority of food supplies bound for Yakutia in winter are delivered from cities such as Khabarovsk and Irkutsk and require ice roads over numerous crossings. Other winter roads follow the river itself, such as the route between Ust'-Kut and Kirensk on the upper Lena River. Resupply occurs during a two- to five-month period determined by weather conditions, ice thickness surveys, and local regulations. In Yakutia, ice roads are normally open from December to April with the northernmost roads closing in early May. After closing, winter roads return to an impassable state and resupply may take place only by air or river barge. Around the Arctic, such roads also serve as access routes for search and rescue teams and recreation-seekers in back-country areas, and provide alternate routes in the event of primary all-weather road closure (Rollheiser 2011).

Use of winter roads depends on subzero temperatures to maintain ground strength and ice thickness. Freezing temperatures must persist long enough to promote buildup of ice to a critical thickness before vehicles may travel safely (ACIA 2004b). Typically, a temperature of -10°C is required before construction may begin (Cardinal 2011). While freezing temperatures are typical throughout the high latitudes in winter, interannual climate variability leads to uncertainty in the length of the winter road operating season each year. For example, a short-term positive-degree warming event, or "Chinook," usually precipitates an early road closure within the following seven to ten days, during which time travel may be possible only at night (Kirschner 2011). In addition to within-season warming events, warmer temperatures can delay the opening and advance the closing date of winter road operation, as has occurred on Alaska's North Slope over a 32-year period (Hinzman et al. 2005). A long-term warming trend therefore implies a future of shorter winter road seasons and reduced terrestrial access in the north.

Stephenson, Smith, and Agnew (2011) modeled the impact of future warming on winter road access by quantifying the change in climatic suitability for building winter roads. Because winter roads may be built relatively inexpensively wherever terrain and climatic conditions allow, mapping suitability rather than actual location of winter roads permits projections of potential inland access rather than operating season lengths of specific winter roads. Relative to a present-day baseline scenario, the authors found a broad pattern of winter road suitability loss from October to May in all Arctic states by midcentury, particularly in Russia and Canada where winter road-suitable land area is greatest (Table 8.1). Losses are especially pronounced in April and November, months representing the "margins" of the operating season in which winter roads typically open and close, respectively. These projections suggest longer travel times near Arctic settlements, especially when combined with rugged terrain and infrastructure scarcity. Travel time to the nearest settlement in parts of Siberia and the Far East is projected to increase by as much as several days by midcentury during the months of greatest winter road loss (April/November). However, decreases in travel time were projected in limited areas during fall months due to delayed freeze-up of navigable rivers. This suggests that inland access change due to climate warming will be complex and region specific, with riverine barge transport increasing in tandem with winter road decline.

Reduced winter road access has numerous negative consequences for residents of northern cities. Often, winter roads are the only land links around and between remote settlements, such as Olenek and Mirnyi in northwestern Yakutia (Tananaev 2013). A shortened winter road season implies higher costs in these impacted areas. When winter roads fail, overland travel may become dangerous or impossible, as drivers may become trapped in muskeg or fall through river or lake ice. Unless located near a navigable waterway, communities face steep price increases when winter roads close as supplies must be delivered by air. In northern Canada, high prices have impelled some residents to ignore safety warnings and drive on thawing winter roads (Champagne 2011). For low-income aboriginal communities already reliant on the state for critical services, reduced mobility represents an additional hardship.

Likewise, mining, energy, and timber interests face shorter time windows to transport necessary equipment and product. This reduced access may lessen the attractiveness of exploitation in remote areas, potentially requiring costly investment in permanent roads and/or maritime port facilities. A recent study of mining in the Northwest Territories shows that failed ice roads may cause annual supply delivery losses of up to $84 million (Goldstein 2011), indicating that reliance on winter roads carries significant economic risk. Furthermore, interannual variability in winter road season length may complicate supply chain coordination, leading to production inefficiencies that may render industrial activities uneconomic. Thus, companies that use winter roads must be able to adapt to both interannual variability in season length in the short term and a truncated average season in the long term.

One solution to the problem of reduced winter road access may be wood fiber roads, a hybrid alternative to permanent and winter roads currently being developed by Shell Canada (Zaitsoff 2011). Fiber roads are composed of a layer of wood chips laid over frozen soil. The fibers are designed to retain high structural integrity when stacked vertically in a layer 0.3–1 meter thick, insulating the ground to prevent thaw and provide a firm driving surface year-round. Shell has deployed fiber roads at several in-situ extraction sites near Peace River, Alberta, in order to test their potential as a replacement for permanent and traditional winter roads. The roads remain functional four years after their construction and currently provide public access for recreational hunting. This longevity makes it possible to amortize the construction cost over several seasons of operation. Even so, fiber roads can be several times more expensive than traditional winter roads (though cheaper than permanent roads), depending on local land features, availability of fiber stock, and regulatory standards. This additional cost may be prohibitive for smaller developers and local governments. Because fiber road development is currently at an experimental stage, more research is needed to establish construction standards, usage loads, and expected longevity. In addition, the effect of high vehicle speed (greater than 30 km/hr) has not been studied. In summer, when muskeg at the edge of fiber roads has thawed, high speeds may cause road damage and dangerous "blowouts" generated by subsurface force waves. If safety standards can be established and construction costs lowered, fiber roads may become a viable year-round alternative to traditional winter roads.

Infrastructure and Future Development

Given the climatic and economic uncertainties, one might envision several contrasting scenarios of future development in Arctic Russia. In one scenario, persistent and unpredictable ice conditions and rising global oil and gas supplies stymie the creation of an Arctic marine transit network based on the NSR. An opposing scenario sees rapid decline in ice extent and attendant expansion of the navigation season combined with high oil prices precipitating large investments in infrastructure and Arctic ports, enabling rapid oil and gas development. In this latter scenario, investment will occur where economic incentives are present and where new infrastructure complements existing facilities and supply lines in fulfilling a strategic or commercial objective (Brigham 2013). It is thus plausible that Naryan Mar, situated on the seasonally ice-free Pechora Sea with close proximity to the Varandey oil terminal and Prirazlomnoye offshore oil field, will be the target of future investment while Tiksi, located far from planned hydrocarbon projects and potential NSR transshipment hubs, will not. Whereas NSR infrastructure in the Soviet period was developed to be independent of international commodity flows, new infrastructure will be built to capitalize on opportunities created by these same flows. Russian Arctic development will thus proceed as a series of targeted, concentrated investments that favor coastal settlements in areas of high resource potential. It is unlikely that inland cities connected by road and rail will benefit from increased investment in the NSR.

History suggests that national development priorities, rather than retreating sea ice, will continue to be a primary driver of these investments. During the Soviet period, establishing an Arctic marine transport system via the NSR was a key strategic aim in support of Siberian natural resource development (Brigham 1991). Traffic along the NSR peaked in 1987 with 6,579 million tons of cargo carried by 331 ships on 1,306 voyages (Brigham and Ellis 2004). Most of these voyages were internal, with first year-round operations beginning in 1978 between Murmansk and Dudinka on the Yenisey River. The route allowed uninterrupted transport between nickel mines at Norilsk and smelters on the Kola Peninsula, and was facilitated by a large fleet of nuclear and diesel-electric powered icebreakers escorting ice-strengthened carriers in convoy. In this context, recent transit activity totaling 46 voyages in 2012 and 71 in 2013 appears modest, while the remarkable growth in activity from 1950–1990 took place absent of any substantive discussion of climate change or receding ice. As in the Soviet era, current resource projects are being driven by government policy and investment support, exemplified by the rapid construction of the Yamal LNG export facility and adjacent port at Sabetta. Even though ice is present most of the year in the Ob Bay, construction is ahead of schedule due to government investments exceeding 47 billion rubles, aimed at doubling current LNG output nationwide and driven by high expected demand in East Asia (Staalesen 2012). The government cemented its support for the project by liberalizing LNG exports beginning in 2014, ending Gazprom's long-standing monopoly on gas exports. Such decisions are ultimately the primary determinants of future shipping activity, reducing the role of sea ice to that of a mitigating factor.

If shipping related to both transit and petroleum activities continues apace, substantial infrastructural upgrades will be required throughout the NSR. Many ports and support infrastructure along the NSR fell into disrepair following the Soviet Union's collapse, and the communities that maintained these facilities experienced severe outmigration (Heleniak 2008). For example, the population of Dikson fell 87 percent from 1989–2010 (CIS Statistical Committee, 1996; *Rossiyskaya Gazeta* 2011). This infrastructure would have to be rebuilt or repaired with a replenished workforce in order to sustain a concentrated expansion of NSR shipping. Such upgrades at ports include deepwater berths, cranes, refueling stations, intermodal transfer facilities, storage tanks and warehouses, worker accommodations, and medical facilities. For transit shipping, ports at the western or eastern terminus of the NSR (for example, Murmansk and Pevek, respectively) will require transshipment capabilities to handle cargo transfer between ordinary freighters and ice-capable vessels. Weather and ice monitoring, firefighting, and rescue stations will be required throughout the NSR to help prevent and respond to emergencies. The Russian government plans to build ten such stations at a cost of 1 billion rubles (Pettersen 2012). Drilling platforms, disaster response vessels, and spill cleanup equipment that operate year-round must be built to withstand ice impacts. Building such infrastructure may require buy-in from foreign companies in the form of capital and expertise. Buoyed by the widely hailed 2010 Barents Sea boundary agreement, Russian and Norwegian companies are signing joint ventures to develop Russian fields in the Barents Sea (Kolyandr 2012). Such partnerships make sense in light of the considerable technical advantages Norway possesses in the oil and gas sector (Stephenson 2012), though other companies, such as Total, may be reluctant to invest in light of Shell's recent difficulties (Russell 2012). By 2015, low oil prices put an end to much of this activity.

Russia will also need to expand its fleet of ice-capable ships. Many of its existing icebreakers face retirement by 2017, and few tankers and cargo vessels in service on the NSR carry Polar Class designation (Vukmanovic and Koranyi 2013). Due to uncertainty in the length of the navigation season, Arctic fleets will need to consist primarily of ice-capable vessels in order to ensure continued operations in colder years. Polar Class ships feature numerous structural improvements over ordinary "open water" ships including strengthened hulls, rudders, and propulsion systems (Brigham, Grishchenko, and Kamisaki 1999; Mulherin, Sodhi, and Smallidge 1994). In addition, insurance costs for trans-Arctic voyages are high owing to environmental hazards and the relative lack of detailed charts, navigational aids, monitoring stations, and port facilities (Tamvakis, Granberg, and Gold 1999; Verny and Grigentin 2009). Local regulations may require ships to have on board a certified "ice navigator" capable of identifying nearby ice regimes and recommending tactical navigation decisions (Østreng 2012; Tamvakis, Granberg, and Gold 1999). Specialized training for other crew members may also be required (Østreng 1999). Icebreaker escort is currently required in the NSR in accordance with UNCLOS Article 234, which grants coastal states the right to "enforce non-discriminatory laws and regulations for the prevention, reduction and control of marine pollution from vessels in ice-covered areas within the limits of the exclusive economic zone." Russian icebreaker fees vary by ship size, cargo type, and the region of the NSR being navigated, but are generally high compared to other cost components (Liu and Kronbak 2010). This fee structure dampens the attractiveness of the NSR regardless of future expansions of the navigation season (Stephenson, Brigham, and Smith 2014).

The need for ice-capable ships highlights the possibility of future tension between the Russian government and private companies operating on the NSR. Mandatory icebreaker escort affords control over the NSR transport sector to Moscow through its state icebreaker company Rosatomflot, as few private shipping companies maintain icebreaker fleets. This allows the government to extract revenue from Arctic shipping while controlling the type and volume of activity in the region. Thus, private companies looking to obtain greater control of their supply chains may find it worthwhile to invest in their own icebreakers, as Norilsk Nickel has done since 2006. As part of an overall process of modernizing its operations, Norilsk Nickel began purchasing new diesel-electric icebreakers to displace the state-owned nuclear icebreakers it had used previously for nickel shipping (Humphreys 2011). Although government intervention has since reduced the degree of autonomy enjoyed by Norilsk Nickel, the icebreaker fleet remains independent, raising the possibility that the company may operate at odds with government interests in the future. If Arctic shipping becomes increasingly economically attractive, shipping companies both within and outside the Arctic will consider investing in ice-capable fleets, weakening Moscow's influence in the region. In this way, efforts to develop Arctic shipping that initially included direct government control of transport links may evolve toward a future of greater independent private operation, forcing the government to reconsider its role as facilitator and overseer of development.

Conclusion

Transportation systems in the Arctic will undergo profound changes in the coming decades. Observed and projected trends in sea ice recession have signaled that marine access will continue to rise throughout the Arctic for most vessel types in summer, while vessels with some icebreaking capability will have expanded access year-round. However, considerable uncertainties remain about the economic impacts of these changes. Resource extraction and transit shipping are risky and expensive enterprises in the Arctic, owing to high infrastructural costs, commodity price fluctuations, navigation season uncertainty, and potentially severe environmental impacts. Furthermore, warming presents an economic vulnerability to inland northern communities and industry through reduced terrestrial winter road accessibility. Resource development and shipping will require considerable investments in infrastructure in order to operate safely and adapt to a changing global economy. In particular, transit shipping will require significant overhaul of port facilities to accommodate cargo transfer between ordinary freighters and ice-capable vessels. While shipping is likely to continue to increase in the near term, it may be many years before Arctic cities become fully integrated within global resource and transportation networks. Policies promoting infrastructural investment and international coordination may ultimately play a bigger role in the growth of Arctic shipping than receding ice. Nevertheless, climate change will remain a critical backdrop for the evolution of Arctic development.

Scott R. Stephenson is Assistant Professor in the Department of Geography at the University of Connecticut.

Notes

1. Special thanks to Robert Orttung and Colin Reisser for their invaluable commentary and for inviting my contribution to this volume. Thanks also to John Agnew, Oleg Anisimov, Tatiana Argounova-Low, Lawson Brigham, Stan Champagne, Gertrude Saxinger, Jessica Graybill, Tim Heleniak, Dave Kirschner, Vera Kuklina, Marlene Laruelle, Elena Nuikina, Elisabeth Öfner, Andrey Petrov, David Rigby, Kris Rollheiser, Nikolay Shiklomanov, Laurence Smith, Dmitry Streletskiy, Nikita Tananaev, Elana Wilson Rowe, and Ken Zaitsoff for helpful comments on an earlier version of this manuscript.

References

ACIA. 2004a. "Cryosphere and Hydrology," in *Arctic Climate Impact Assessment.* Cambridge: Cambridge University Press.
———. 2004b. "Infrastructure: Buildings, Support Systems, and Industrial Facilities," in *Arctic Climate Impact Assessment.* Cambridge: Cambridge University Press.
Adam, K.M. 1978. *Building and Operating Winter Roads in Canada and Alaska.* Ottawa: Department of Indian and Northern Affairs.
Agnew, J. 2002. *Place and Politics in Modern Italy.* Chicago: University of Chicago Press.
AMAP. 2007. *Arctic Oil and Gas 2007.* Oslo: Arctic Monitoring and Assessment Programme.
American Petroleum Institute. 2006. *Joint Association Survey on Drilling Costs, 2000 and 2005 Editions.* Washington, DC: American Petroleum Institute.
Arbo, P., et al. 2012. "Arctic Futures: Conceptualizations and Images of a Changing Arctic," *Polar Geography* 36: 163–82.
Arctic Council. 2009. *Arctic Marine Shipping Assessment 2009 Report.* Tromsø, Norway: Arctic Council.
Argounova-Low, T. 2012. "Roads and Roadlessness: Driving Trucks in Siberia," *Journal of Ethnology and Folkloristics* 6: 71–88.
Atlas, R.M., A. Horowitz, and M. Busdosh. 1978. "Prudhoe Crude Oil in Arctic Marine Ice, Water, and Sediment Ecosystems: Degradation and Interactions with Microbial and Benthic Communities," *Journal of the Fisheries Research Board of Canada* 35: 585–90.

Barber, D.G., et al. 2010. "The International Polar Year IPY Circumpolar Flaw Lead CFL System Study: Overview and the Physical System," *Atmosphere-Ocean* 48: 225–43.

Belchansky, G.I., D.C. Douglas, and N.G. Platonov. 2004. "Duration of the Arctic Sea Ice Melt Season: Regional and Interannual Variability, 1979–2001," *Journal of Climate* 17: 67–80.

Bennett, M. 2013. "Arctic Frontiers Conference Gathers the Best and Brightest of the North," *Alaska Dispatch*, 22 January.

BP. 2011. *BP Statistical Review of World Energy*, June.

Brigham, L.W. (ed.). 1991. *The Soviet Maritime Arctic*. Annapolis, MD: Naval Institute Press.

———. 2013. "Arctic Marine Transport Driven by Natural Resource Development," *Baltic Rim Economies Quarterly Review* 2: 13–14.

Brigham, L.W. and B. Ellis (eds). 2004. *Arctic Marine Transport Workshop Report. Scott Polar Research Institute, University of Cambridge, 28–30 September*. Anchorage, AK: Institute of the North.

Brigham, L.W., V.D. Grishchenko, and K. Kamisaki. 1999. "The Natural Environment, Ice Navigation and Ship Technology," in W. Østreng (ed.), *The Natural and Societal Challenges of the Northern Sea Route*. London: Kluwer Academic Publishers.

Broder, J.M. 2013. "Interior Department Warns Shell on Arctic Drilling," *New York Times*, 14 March.

Bryanski, G. 2011. "Russia's Putin Says Arctic Trade Route to Rival Suez," *Reuters*, 22 September.

Budzik, P. 2009. *Arctic Oil and Gas Potential*. Washington, DC: US Energy Information Administration, Office of Integrated Analysis and Forecasting.

Cardinal, M. 2011. Owner/President, Lakeshore Enterprises. Personal communication.

Champagne, S. 2011. Deputy Assessor, Northern Sunrise County. Personal communication.

Christensen, J.H., et al. 2007. "Regional Climate Projections," in S. Solomon et al. (eds), *Climate Change 2007: The Physical Science Basis. Contribution of Working Group I to the Fourth Assessment Report of the Intergovernmental Panel on Climate Change*. Cambridge: Cambridge University Press.

CIS Statistical Committee. 1996. 1989 USSR Census CD-ROM. EastView Publications.

Cole, H., V. Colonell, and D. Esch 1999. *The Economic Impact and Consequences of Global Climate Change on Alaska's Infrastructure*. Fairbanks, AK: US Global Change Research Program, University of Alaska-Fairbanks.

Comiso, J.C. 2012. "Large Decadal Decline of the Arctic Multiyear Ice Cover," *Journal of Climate* 25: 1176–93.

Comiso, J.C., C.L. Parkinson, R. Gersten, and L. Stock 2008. "Accelerated Decline in the Arctic Sea Ice Cover," *Geophysical Research Letters* 35: L01703, doi:10.1029/2007GL031972.

Corbett, J.J., et al. 2010. "Arctic Shipping Emissions Inventories and Future Scenarios," *Atmospheric Chemistry and Physics* doi:10: 9689–9704.

Det Norske Veritas. 2011. *Heavy Fuel in the Arctic Phase 1*. Report No. 2011–0053. PAME.

EIA. 2008. *Analysis of Crude Oil Production in the Arctic National Wildlife Refuge.* Washington, DC: US Energy Information Administration.
———. 2011. *World Oil Transit Chokepoints.* Washington, DC: US Energy Information Administration.
Gautier, D.L., et al. 2009. "Assessment of Undiscovered Oil and Gas in the Arctic," *Science* 324: 1175–79.
Goldstein, M. 2011. "Cold Hard Cash: The Economic Importance of Ice in the Arctic," *7th International Congress of Arctic Social Sciences,* Akureyri, Iceland, 22 June.
Guyer, S. and B. Keating 2005. *The Impact of Ice Roads and Ice Pads on Tundra Ecosystems, National Petroleum Reserve-Alaska.* Washington, DC: US Department of the Interior, Bureau of Land Management.
Heleniak, T. 2008. "Changing Settlement Patterns Across the Russian North at the Turn of the Millennium," in M. Tykkylainen and V. Rautio (eds), *Russia's Northern Regions on the Edge: Communities, Industries and Populations from Murmansk to Magadan.* Helsinki: Kikimora Publications, University of Helsinki.
Hinzman, L.D., et al. 2005. "Evidence and Implications of Recent Climate Change in Northern Alaska and Other Arctic Regions," *Climatic Change* 72: 251–98.
Howell, S. and J. Yackel 2004. "A Vessel Transit Assessment of Sea Ice Variability in the Western Arctic, 1969–2002: Implications for Ship Navigation," *Canadian Journal of Remote Sensing* 30: 205–15.
Hulbert, M. 2012. "A Six Point Plan for Putin Hydrocarbon Heaven," *Forbes,* 26 April.
Humphreys, D. 2011. "Challenges of Transformation: The Case of Norilsk Nickel," *Resources Policy* 36: 142–48.
International Boundaries Research Unit. 2008. *Maritime Jurisdiction and Boundaries in the Arctic Region.* Durham University.
International Energy Agency. 2010. *World Energy Outlook.* Paris: IEA.
IPCC. 2013. "Summary for Policymakers." In Stocker, T.F., D. Qin, G.-K. Plattner, M. Tignor, S. K. Allen, J. Boschung, A. Nauels, Y. Xia, V. Bex and P.M. Midgley (eds.): *Climate Change 2013: The Physical Science Basis. Contribution of Working Group I to the Fifth Assessment Report of the Intergovernmental Panel on Climate Change.* Cambridge and New York: Cambridge University Press.
Johnston, R. 1986. *The Dictionary of Human Geography,* 2nd ed. Oxford: Blackwell.
Joling, D. 2012. "Sparse Data on Ocean Depth Challenges NOAA in Arctic," *Anchorage Daily News,* 23 May.
Khon, V.C., et al. 2010. "Perspectives of Northern Sea Route and Northwest Passage in the Twenty-First Century," *Climatic Change* 100: 757–68.
Kirschner, D. 2011. Council member, Northern Alberta Development Council. Personal communication.
Kolyandr, A. 2012. "Norway's Statoil Signs Arctic Deal with Russia's Rosneft," *Wall Street Journal,* 6 May.
Kroh, K. 2012. "German Bank Won't Finance Arctic Ocean Drilling, Saying the 'Risks and Costs Are Simply Too High'," *ThinkProgress.org,* 23 April.
Kwok, R., et al. 2009. "Thinning and Volume Loss of the Arctic Ocean Sea Ice Cover: 2003–2008," *Journal of Geophysical Research* 114: C07005.
Larsen, P.H., et al. 2008. "Estimating Future Costs for Alaska Public Infrastructure at Risk from Climate Change," *Global Environmental Change* 18: 442–57.

Laxon, S., N. Peacock, and D. Smith. 2003. "High Interannual Variability of Sea Ice Thickness in the Arctic Region," *Nature* 425: 947–50.

Lindholt, L. and S. Glomsrød. 2012. "The Arctic: No Big Bonanza for the Global Petroleum Industry," *Energy Economics* 34: 1465–74.

Liu, M. and J. Kronbak 2010. "The Potential Economic Viability of Using the Northern Sea Route NSR as an Alternative Route between Asia and Europe," *Journal of Transport Geography* 18: 434–44.

Lloyd's. 2012. *Arctic Opening: Opportunity and Risk in the High North.* London: Chatham House.

Martin, S. 1979. "A Field Study of Brine Drainage and Oil Entrainment in First-Year Sea Ice," *Journal of Glaciology* 22: 473–502.

Maslanik, J., J. Stroeve, C. Fowler, and W. Emery. 2011. "Distribution and Trends in Arctic Sea Ice Age through Spring 2011," *Geophysical Research Letters* 38: L13502, doi:10.1029/2011GL047735.

Maslanik, J.A., et al. 2007. "A Younger, Thinner Arctic Ice Cover: Increased Potential for Rapid, Extensive Sea-Ice Loss," *Geophysical Research Letters* 34: L24501, doi:10.1029/2007GL032043.

McGrath, M. 2012. "Gas Tanker Ob River Attempts First Winter Arctic Crossing." BBC, 25 November.

Melling, H. 2002. "Sea Ice of the Northern Canadian Arctic Archipelago," *Journal of Geophysical Research* 107: 3181.

Moe, A. and O. Jensen 2010. *Opening of New Arctic Shipping Routes.* Directorate-General for External Policies, European Parliament. PE 433.792.

Mulherin, N.D., et al. 1996. *Development and Results of a Northern Sea Route Transit Model.* Hanover, NH: US Army Corps of Engineers CRREL.

Mulherin, N.D., D. Sodhi, and E. Smallidge. 1994. *Northern Sea Route and Icebreaking Technology: An Overview of Current Conditions.* Hanover, NH: US Army Corps of Engineers CRREL.

National Oil Spill Commission. 2012. *Assessing Progress: Implementing the Recommendations of the National Oil Spill Commission.* Washington, DC: National Oil Spill Commission.

NSIDC. 2012a. "Arctic Sea Ice Extent Breaks 2007 Record Low." Retrieved from http://nsidc.org/arcticseaicenews/2012/08/arctic-sea-ice-breaks-2007-record-extent/.

———. 2012b. "Arctic Sea Ice Maximum Marks Beginning of Melt Season." Retrieved from http://nsidc.org/arcticseaicenews/2012/03/arctic-sea-ice-maximum-marks-beginning-of-melt-season/.

NSR Information Office. 2013. "Final Statistics Figures for Transit Navigation on the NSR in 2013." Retrieved from http://arctic-lio.com/docs/nsr/transits/Transits_2013_final.pdf.

Oil and Energy Trends. 2012. "FOCUS: Is Russia's Oil Production About to Decline?" *Oil and Energy Trends* 37: 3–6.

Østreng, W. (ed.). 1999. *The Natural and Societal Challenges of the Northern Sea Route.* London: Kluwer Academic Publishers.

———. 2012. *Shipping in Arctic Waters.* Kirkenes, Norway: Center for High North Logistics.

Parry, M.L., et al. 2007. *Impacts, Adaptation and Vulnerability: Contribution of*

Working Group II to the Fourth Assessment Report of the Intergovernmental Panel on Climate Change. Cambridge: Cambridge University Press.

Paxian, A., et al. 2010. "Present-Day and Future Global Bottom-Up Ship Emission Inventories Including Polar Routes," *Environmental Science and Technology* 44: 1333–39.

Peters, G.P., et al. 2011. "Future Emissions from Shipping and Petroleum Activities in the Arctic," *Atmospheric Chemistry and Physics* 11: 5305–20.

Pettersen, T. 2011a. "First Supertanker along Northern Sea Route," *Barents Observer*, 24 August.

———. 2011b. "Speed Record on Northern Sea Route," *Barents Observer*, 17 August.

———. 2012. "Russia to Install Naval Facilities in the Arctic," *Barents Observer*, 7 August.

Pew Environment Group. 2010. *Oil Spill Prevention and Response in the U.S. Arctic Ocean: Unexamined Risks, Unacceptable Consequences*. Philadelphia: Pew Environment Group.

Polyakov, I.V., J.E. Walsh, and R. Kwok. 2012. "Recent Changes of Arctic Multiyear Sea Ice Coverage and the Likely Causes," *Bulletin of the American Meteorological Society*, DOI:10.1175/BAMS-D-11-00070.1.

Rollheiser, K. 2011. Senior Development Officer, Northern Alberta Development Council. Personal communication.

Rosstat. 2011. 2010 Russian Census. Rossiiskaia Gazeta.

Russell, J. 2012. "Total Warns Against Drilling for Oil in Arctic," *The Telegraph*, 26 September.

Serreze, M.C., M.M. Holland, and J.C. Stroeve. 2007. "Perspectives on the Arctic's Shrinking Sea-Ice Cover," *Science* 315: 1533–36.

Smith, L.C. and S.R. Stephenson. 2013. "New Trans-Arctic Shipping Routes Navigable by Midcentury," *Proceedings of the National Academy of Sciences* 110: 4871–72.

Staalesen, A. 2012. "Yamal LNG Ahead of Schedule," *Barents Observer*, 3 December.

———. 2013. "First Oil from Yamal," *Barents Observer*, 15 May.

Stephenson, S.R. 2012. "Collaborative Infrastructures: A Roadmap for International Cooperation in the Arctic," *Arctic Yearbook 2012* 1: 311–33.

Stephenson, S.R. and J.A. Agnew. 2016. "The Work of Networks: Embedding Firms, Transport and the State in the Russian Arctic Oil and Gas Sector," *Environment and Planning A* 48: 558-576.

Stephenson, S.R., L.W. Brigham, and L.C. Smith 2014. "Marine Accessibility Along Russia's Northern Sea Route," *Polar Geography* 37: 111–33.

Stephenson, S.R. and L.C. Smith. 2015. "Influence of Climate Model Variability on Projected Arctic Shipping Futures," *Earth's Future* 3: 331-343.

Stephenson, S.R., L.C. Smith, and J.A. Agnew. 2011. "Divergent Long-Term Trajectories of Human Access to the Arctic," *Nature Climate Change* 1: 156–60.

Stephenson, S.R., L.C. Smith, L.W. Brigham, and J.A. Agnew. 2013. "Projected 21st-Century Changes to Arctic Marine Access," *Climatic Change* 118: 885–99.

Streletskiy, D.A., N. Shiklomanov, and F.E. Nelson. 2012a. "Permafrost, Infrastructure, and Climate Change: A GIS-based Landscape Approach to Geotechnical Modeling," *Arctic, Antarctic, and Alpine Research* 44: 368–80.

———. 2012b. "Spatial Variability of Permafrost Active-Layer Thickness under Contemporary and Projected Climate in Northern Alaska," *Polar Geography*. DOI:10.1080/1088937X.2012.680204.
Stroeve, J.C., et al. 2008. "Arctic Sea Ice Extent Plummets in 2007," *Eos, Transactions American Geophysical Union* 89: 13–20.
———. 2012a. "Trends in Arctic Sea Ice Extent from CMIP5, CMIP3 and Observations," *Geophysical Research Letters* 39: L16502.
———. 2012b. "The Arctic's Rapidly Shrinking Sea Ice Cover: A Research Synthesis," *Climatic Change* 110: 1005–27.
Taaffe, E.J. and H.L. Gauthier. 1973. *Geography of Transportation*. Englewood Cliffs, NJ: Prentice-Hall.
Tamvakis, M., A.G. Granberg, and E. Gold. 1999. "Economy and Commercial Viability," in W. Østreng (ed.), *The Natural and Societal Challenges of the Northern Sea Route*. London: Kluwer Academic Publishers.
Tananaev, N.I. 2013. Head, Igarka Geocryology Lab, Permafrost Institute, Yakutsk Science Center, Siberian Division, Russian Academy of Science. Personal communication.
US Government Accountability Office. 2012. *Oil and Gas: Interior Has Strengthened Its Oversight of Subsea Well Containment, but Should Improve Its Documentation*. Report No. GAO-12-244. Washington, DC: US Government Accountability Office.
USARC Permafrost Task Force. 2003. *Climate Change, Permafrost, and Impacts on Civil Infrastructure*. Special Report 01–03. Arlington, VA: US Arctic Research Commission.
Vavrus, S.J., et al. 2012. "Twenty-First-Century Arctic Climate Change in CCSM4," *Journal of Climate* 25: 2696–710.
Verma, M.K., L.P. White, and D.L. Gautier. 2008. *Engineering and Economics of the USGS Circum-Arctic Oil and Gas Resource Appraisal CARA Project*. USGS. Open-File Report 2008–1193.
Vermaire, J.C., et al. 2013. "Arctic Climate Warming and Sea Ice Declines Lead to Increased Storm Surge Activity," *Geophysical Research Letters* 40: 1386–90.
Verny, J. and C. Grigentin. 2009. "Container Shipping on the Northern Sea Route," *International Journal of Production Economics* 122: 107–17.
Vukmanovic, O. and B. Koranyi. 2013. "Russia's Revival of Arctic Northern Sea Route at Least 10 Years Away," *Reuters*, 25 January.
Wang, M. and J.E. Overland. 2009. "A Sea Ice Free Summer Arctic Within 30 Years?" *Geophysical Research Letters* 36: L07502.
Westbrook, J. 2012. "Russia Oil Output Relies on Foreign Investment, Sechin Tells WSJ," *Bloomberg*, 22 April.
Yergin, D. 2012. *The Quest: Energy, Security, and the Remaking of the Modern World*. New York: Penguin.
Zaitsoff, K. 2011. Consultation Coordinator, Shell Canada. Personal communication.
Zhang, X. and J.E. Walsh. 2006. "Toward a Seasonally Ice-Covered Arctic Ocean: Scenarios from the IPCC AR4 Model Simulations," *Journal of Climate* 19: 1730–47.

CHAPTER NINE

Russian Arctic Cities through the Prism of Permafrost

Dmitry Streletskiy and Nikolay Shiklomanov

Introduction

While the Arctic regions are generally sparsely populated, the population of the Russian Arctic is relatively large.[1] Urbanization in most of the Russian Arctic regions is higher than overall in Russia with the majority of the population concentrated in settlements with a distinct urban fabric (Rosstat 2014; Map 9.1). There are numerous small towns scattered throughout the Russian Arctic, however, substantial populations are concentrated in large urban centers such as Vorkuta, Salekhard, Nadym, Novyy Urengoy, Norilsk, Magadan, and Yakutsk. Even though high urbanization numbers are characteristic of the mainly non-indigenous population, many peoples of the North are integrated into the urban areas as well (Rosstat 2014). Although only about 5 percent of the total Russian population resides in Arctic regions, their contribution to the country's economy is disproportionally large. Therefore, maintaining labor resources and infrastructure in the Arctic is one of the key requirements for the continued prosperity of Russia's resource-oriented economy.

Ongoing and projected climatic warming presents significant challenges in sustaining Arctic communities as most of them are built on permafrost. Permafrost, defined as any subsurface material that remains frozen for more than two consecutive years, is widespread in northern latitudes, and almost two-thirds of Russian territory is affected by this phenomenon (Zhang et al. 2008). Permafrost deter-

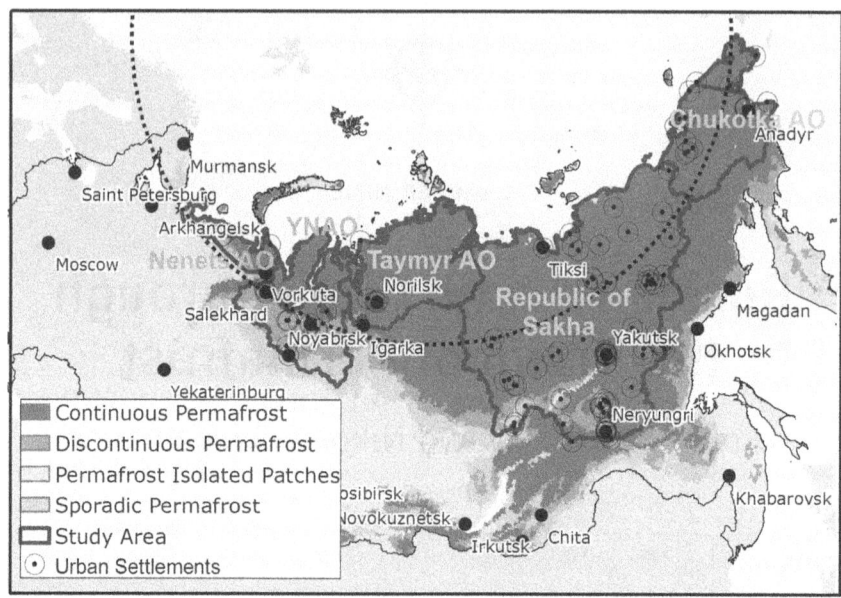

Map 9.1. | *Study Area and Location of Urban Settlements in Russia Relative to Permafrost Distribution*

mines many important environmental processes, and presents numerous challenges to human development in the Arctic.

Almost all types of human activities involve significant alteration of the natural ground covers, promoting warming and degradation of permafrost. In turn, the bearing capacity of permafrost decreases with its warming, causing the weakening of foundations and potential damage to, and possible failure of, buildings, pipelines, and transportation facilities (Streletskiy, Anisimov, and Vasiliev 2014). To address these challenges, engineers utilize a range of practices to construct and maintain infrastructure on permafrost, making it possible to develop large tracts of the Arctic.

Climate warming has the potential to further reduce soil bearing strength, increase soil permeability, and intensify the potential for the development of such cryogenic processes as differential thaw settlement and heave, destructive mass movements, and the development of thermokarst terrain (Shiklomanov and Nelson 2013). Each of these phenomena has the capacity to cause severe negative consequences for urban infrastructure in the high latitudes (Khrustalev and Davidova 2007; Anisimov and Streletskiy, 2015). The difficulty of relocating urban settlements requires developing an adaptive capacity to mitigate the detrimental, permafrost-related impacts of climatic changes.

Numerous buildings in Russian Arctic settlements have experienced substantial numbers of structural deformations (Kronic 2001; Grebenets, Streletskiy, and Shiklomanov 2012). In many cases, influences such as age, lack of maintenance, or design/construction flaws affect a structure. However, recent studies (Khrustalev and Davidova 2007; Khrustalev, Parmuzin, and Emelyanova 2011; Streletskiy, Shiklomanov, and Hateberg 2012; Anisimov and Streletskiy 2015; Streletskiy et al. 2015) indicate that ongoing changes in climatic conditions partially explain the intensified rate of structural deformations of buildings erected on permafrost. Transportation routes and facilities are also subject to danger (Kondratiev 2013). The long lateral extent of this type of infrastructure makes it difficult to choose an optimum route and apply economically sound strategies for controlling cryogenic processes. Warmer air temperatures further limit the accessibility of remote regions due to winter road deterioration, as explained in Chapter 8 (Stephenson, Smith, and Agnew 2011).

As climate warming continues to evolve (for details consult Chapter 7), it may further intensify detrimental impacts on infrastructure throughout the permafrost regions, even under appropriate environmental management and engineering practices. Collapsing roads and buildings can have severe socioeconomic consequences, since supporting most of the existing infrastructure will require expensive mitigation strategies and the associated economic impacts may reach far beyond the Arctic (Whiteman, Hope, and Wadhams 2013). This chapter attempts to quantitatively evaluate climate and technogenic impacts facing large settlements and their residents in the Russian Arctic under rapidly changing climatic conditions.

Permafrost as an Integrative Component of Climate- and Human-Induced Change

With permafrost occupying about a quarter of the Northern Hemisphere's land surface, Canada and Russia contain the most extensive areas of permafrost: approximately 50 percent and 65 percent of their territories, respectively. The thickness of permafrost varies from a few meters near its southern limit to several hundred meters in the high Arctic. The maximum reported thickness is more than 1.4 km in unglaciated parts of Siberia (Washburn 1980).

Even though permafrost is ground that remains below 0°C for more than two years, even in the coldest conditions there is an active layer in which heat flux creates a stratum of seasonal thawing and freezing.

At any given location, the active layer can be defined as the maximum depth of annual thaw penetration. The influence of atmospheric processes on ground thermal conditions is moderated by local hydrology and processes occurring in the boundary layer of vegetation, snow, and surface organic matter (Streletskiy, Shiklomanov, and Nelson 2012b). In addition to climatic changes, any natural and/or anthropogenic disturbances at the ground surface that upsets the delicate equilibrium between climatic and ground thermal regimes are likely to promote permafrost warming, an increase in seasonal thaw propagation, and possibly permafrost degradation (Streletskiy, Anisimov, and Vasiliev 2014). As such, observable changes in permafrost conditions can serve as an integrative indicator of climate- and human-induced environmental change.

At small geographical scales, the distribution of permafrost is classified primarily on the basis of its lateral continuity into continuous, discontinuous, and sporadic zones. Because permafrost is a climatologically determined phenomenon, the distribution of these zones is crudely correlated with climatic zones. Within each zone, the permafrost continuity and thickness are generally increasing in the northward direction, while permafrost temperature and the depth of annual thaw are decreasing. However, this broad zonal pattern is greatly mediated by topographic, microclimatic, surface, and subsurface conditions resulting in a high spatial heterogeneity in ground thermal conditions, even within a small area characterized by uniform climate (Streletskiy, Shiklomanov, and Nelson 2012b).

Many engineering problems related to permafrost construction are associated with the loss of soil bearing capacity and increased potential for the development of such cryogenic processes as differential thaw settlement and heave. These processes usually result from changes in the permafrost thermal regime due to the disturbance of the ground surface, heat propagation from structures, and/or climatic changes (Grebenets, Streletskiy, and Shiklomanov 2012). Soil bearing capacity in permafrost regions is an important problem associated with climate change, primarily because of its strong dependency on temperature. Frozen soil can experience a significant loss of strength when warmed. Frozen soils containing excess ice (ice-rich soils) can be susceptible to settlement and thermokarst development, which occurs when ice-rich soil thaws. The presence of ice in the ground can be attributed to several thermodynamic, geologic, and geomorphologic factors, resulting in uneven vertical and lateral distribution and great variability in its amount and properties (Shiklomanov and Nelson 2013).

Observed climatic warming has resulted in a permafrost temperature increase of 0.5–2°C over the last three decades in the Russian Arctic (Romanovsky et al. 2010). The southern permafrost boundary has retreated northward in the Northern European part of Russia (Oberman 2008). Thawing of ice-rich permafrost is also occurring along the southern fringes of the permafrost zone in West Siberia (Vasiliev, Leibman, and Moskalenko 2008). Previously stable areas of North-Western Yakutia have begun to show evidence of permafrost warming in recent years (Romanovsky et al. 2010). The increase in permafrost temperature has been accompanied by an increase of active layer thickness (ALT) in the majority of the regions, and is especially pronounced in the Russian European North (Shiklomanov, Streletskiy, and Nelson 2012; Romanovsky et al. 2015). Increasing active layer thickness is generally considered to be one of the most immediate reactions of permafrost to the warming climate.

The reported changes in permafrost and the active layer are exacerbated in areas of concentrated human activities, especially in large settlements. The development of previously undisturbed areas is commonly accompanied by the removal of vegetation, changes to surface and subsurface hydrology, redistribution of snow, and modification of the soil's thermal properties. These, in turn, upset the heat exchange balance between the atmosphere and the ground. The resulting technogenically modified complexes are characterized by a suite of permafrost-related processes that are often drastically different from those that were characteristic of the area prior to development (Grebenets, Streletskiy, and Shiklomanov 2012). As such, cities represent a nucleus of anthropogenic impacts on the fragile Arctic environment where climate-induced impacts on permafrost conditions are greatly amplified. The combined human and climate effects on frozen ground for any given settlement depends on numerous factors, including the climate and environmental conditions at the location of the settlement, its size and population density, planning, architectural and engineering practices, and history of development.

The Permafrost Conquest: Construction Practices and Legacies

Russia has a long history of Arctic exploration, which led to the extensive development of its northern territories. For example, the city of Yakutsk, located in a zone of relatively cold, thick continuous permafrost, was founded in 1632. Many permafrost-related challenges arose during the construction of the Trans-Siberian Railway (1891–1916), substantial portions of which lie on permafrost (Shiklomanov 2005). Many engineering practices developed by trial and error during that period are summarized in the first permafrost engineering book titled *Permafrost and Construction upon It* (Bogdanov 1912). Historically, Siberian settlements were characterized by low population densities, marginal environmental impacts, and poor infrastructure since they consisted primarily of small wooden houses on traditional basement foundations. Inexpensive reconstruction or relocation of the light structures to a new location addressed the inevitable deformations that took place.

This situation began to change drastically just before and, especially, after World War II. The vulnerability of the European part of the country to military invasion and the discovery of major reserves of vital mineral resources promoted intensified industrial development in the permafrost-heavy North-Eastern regions. Stalin's decision to concentrate forced labor into gulag camps during the 1930s and 1940s gave him a reliable supply of manual labor to conquer the harsh permafrost environment. For example, prisoners built the first brick and concrete buildings in the city of Norilsk by digging through perennially frozen sediments to the solid bedrock and then building foundations using conventional construction practices. While this approach can erect stable structures, it is prohibitively expensive without low-cost prison labor and applicable only in areas with thin sedimentary layers.

The abolishment of the gulag system in 1953, postwar housing crisis, and development of the North-Eastern regions forced the Soviet Union to adopt construction techniques that were affordable and quick. In the late 1950s, Nikita Khrushchev started a new era of massive construction that relied on cheap, standardized, prefabricated panel buildings. More than 50 million people (or a quarter of the USSR population) moved to new apartments during the first decade of this program. Simultaneously, the Norilsk engineer Mikhail Kim developed piling foundations for permafrost construction, which effectively elevate buildings above the ground surface to avoid heat penetration into the ground and redistribute pressure between several piles to support a large structural load. This construction method also decreased the disturbance of permafrost during construction, as the pile diameter was relatively small compared to other types of foundations. The elevated first floor with a crawl basement was clearly advantageous, as the basement remains ventilated during the winter while being shaded during the summer, preventing the permafrost from thawing.

The introduction of relatively cheap piling foundations effectively conquered permafrost and opened an entirely new chapter in the Soviet history of development in northern regions. The cheap, prefabricated building designs resulted in a construction boom and promoted urbanization throughout the Arctic.

As a result, urban architecture in the Russian Arctic mainly consists of a mixture of five- to nine-floor buildings made from prefabricated panel or standard-design brick. The majority of such structures on permafrost are built using piling foundations, known as the "Passive Principle" of permafrost construction (Shur and Goering 2009) or as "Principle One" in the Russian literature. This approach uses permafrost as the base for the building foundation, protecting it from warming and thawing during construction and the lifespan of the structure, which is engineered to last thirty to fifty years.

The ability of pile foundations to carry a building's structural load depends on many factors, such as ground temperature, texture, density, salinity, ice content, and the presence of unfrozen water. In unfrozen coarse sediments (gravels and denser), soil strength is primarily a function of internal friction. In frozen soils, however, it is a function of ice bonding, allowing for much heavier loads compared to those in non-permafrost regions, as considerable bearing capacity can be gained by freezing pile sides into the surrounding permafrost (*adfreezing*). The shear stress per side unit of a pile is much lower compared to the normal stress at the bottom of a pile, but the much larger adfreezing area of the pile side relative to the bottom allows redistribution of up to 80 percent of the pile bearing capacity to its sides. Adfreezing strength is strongly dependent on the permafrost temperature along the piles, which consequently makes a foundation's total bearing capacity highly dependent on permafrost temperatures, which in turn depend on the climate and insulating properties of ground covers, such as snow and vegetation. Temporal variability of air temperature and snow results in climate-induced changes in the load-bearing capacity of permafrost (Streletskiy, Shiklomanov, and Nelson 2012a). Russian Construction Norms and Regulations on Permafrost (CNR) (CNR 1990) recommend using standard climate statistics available at the time of construction to account for average climatic conditions and possible natural variability. Observed climate changes now, however, might exceed the climate variability engineered into piling foundations constructed during the Soviet era, which can potentially reduce the stability of structures.

Quantifying Infrastructure Stability under Climate Change

Every structure requires an individual engineering approach; however, simple quantitative assessments of foundation stability are possible even at larger, regional scales. To assess changes in the potential stability of Arctic infrastructure, we have used the bearing capacity of the standard 0.35x0.35x10 m concrete foundation pile. Such an approach is frequently used in Russia for preliminary engineering assessments of large territories. This method allows avoiding uncertainties due to diversity of possible construction designs and practices, while focusing on changes in engineering properties of permafrost characteristics in response to climatic forcing. Foundation bearing capacity was calculated using parameterizations provided by the Construction Norms and Regulations (CNR 1990), which relate active layer thickness and permafrost temperature to stresses experienced by the pile foundation imbedded into the permafrost. Changes in permafrost parameters (for example, active layer thickness, permafrost temperature) were estimated using a spatially distributed model of permafrost-climate interactions (Streletskiy, Shiklomanov, and Nelson 2012b).

To evaluate changes in bearing capacity over large geographic areas, the model was developed with climatic inputs obtained from the National Center for Environmental Prediction (Kalnay et al. 1996). The climatic inputs consist of gridded datasets of daily temperatures and precipitation, scaled to 25 km^2 resolution. The assessments were focused on two reference periods: 1965–1975, and 1995–2005. The decadal periods were chosen to illuminate interannual variability in data and to focus on climatic changes rather than on weather oscillations of any particular year. The decade of 1965–1975 was chosen as a baseline period due to the extensive construction in the Russian Arctic at that time. It is assumed that structures were designed to withstand the climatic conditions of the period. The 1995–2005 period was selected to represent contemporary conditions, as at the time of analysis, the most recent climate data were available through 2005. Land surface data is composed of a uniform sand profile with low ice content, which is considered as the most favorable lithological conditions for construction. All the results presented in this chapter are in the form of relative change between the two selected periods.

Geographically, this analysis focuses on five Russian administrative regions bordering the Arctic Ocean and largely covered by permafrost: the Nenets Autonomous Okrug (AO), the Yamal-Nenets AO, the Taymyrsky Dolgano-Nenetsky District (formerly the Taymyr AO), the

Republic of Sakha (Yakutia), and the Chukotka AO. Collectively these regions comprise a significant portion of the Russian Arctic and its population.

To estimate the impacts of climate change on the population, population data were taken from the 2002 Russian Census as it centers on the 1995–2005 period (Rosstat 2014). Some regions experienced population change between the 2002 and 2010 censuses, which is discussed in the following section. However, it is important to note that the 2002 census population should be utilized, as representative of the selected climatic period.

Geography of Russian Arctic Urban Population

A majority of the Russian Arctic population is concentrated in urban centers. While the proportion of the indigenous peoples of the North living in urban areas is less relative to non-indigenous groups, about 30 percent are urban (Rosstat 2014). City residents made up 73 percent of the total 1.85 million people living across the study area in 2001. Although the population in this area decreased to 1.62 million by 2010, the percent of urban population remained the same (Rosstat 2014). One hundred settlements across the five study region accounted for an urban population of 1.36 million people. One-third of the settlements had 2,000–10,000 people, about a fifth had a population of 10,000–50,000 people and six cities were above 50,000.

Sakha Republic had the highest number of settlements and largest population of the five administrative regions within the study area. The 73 settlements located within Sakha accounted for a population of 623,200 (by 2010, the population decreased by only one thousand). This is nearly 65 percent of the study area's total settlements and just under half of the total urban population.

With fourteen settlements, the Chukotka AO had the second greatest number of settlements, though these settlements tend to be much smaller. The urban population of Chukotka in 2001 was 46,000, or less than 5 percent of the study area's total. The largest settlement in Chukotka was Anadyr, with a population of 11,300. Even though Anadyr's population did not change, the number of urban settlements dropped nearly 50 percent by 2010 with only 32,300 people left in the urban settlements. The Yamal-Nenets AO was similar to Chukotka in number of settlements; however, its urban population was the second largest in the study area, accounting for nearly 30 percent of the total urban population with 386,600 people (the urban population increased substantially by 2010 to 464,100). The Taymyrsky Dolgano-Nenetsky District (hereafter Taymyr) had five settlements, with a total population of 252,400. This is 5 percent of the study area's settlements and nearly 20 percent of the population. Geographically Taymyr includes settlements of the Taymyr District (24,000) and the Norilsk Urban Okrug, which are under the jurisdiction of Krasnoyarsk Kray. Dudinka, Talnakh, and Norilsk are the biggest settlements that are within the region's jurisdiction.

The Nenets AO had the fewest settlements and the smallest urban population, with three settlements accounting for 27,200 people (no change in 2010). The Nenets AO accounts for less than 5 percent of the study area's total settlements and urban population.

The two largest settlements of the study area were the cities of Yakutsk in the Sakha Republic and Norilsk in Taymyr. Other cities with populations above 50,000 included: Noyabrsk and Novyy Urengoy in the Yamal-Nenets AO, Neryungri in Sakha, and Talnakh in Taymyr (Talnakh was recently merged into the Norilsk Industrial Okrug).

From 2001 to 2010, Chukotka AO lost almost 30 percent of its population, while Yamal-Nenets AO was able to grow significantly (see Chapter 4). Settlements in the Russian North remain highly concentrated. The indigenous population remained integrated into the urban settlements: 21 percent Nenets, 24 present Chukchi, 38 percent Khanti and as many as 57 percent Mansi were urban, with a much higher percent of women relative to men (Rosstat 2014). Throughout the chapter, Norilsk, Yakutsk, Neryungri, Novyy Urengoy, and Noyabrsk will be referenced as changes in infrastructure parameters in these places could pose consequences to a large number of people.

Effects of Climate Change on the Population

Most of the study area experienced pronounced climate warming between the 1970s and 2000s time periods, with only a few areas showing no warming trend such as the western part of the Nenets AO and Northern Sakha. The mean annual air temperature across the region for the 2000s time period was −9.5°C, up from −10.4°C for the 1970s time period. A majority of settlements experienced temperature increases between 0.5°C and 1.5°C (Map 9.2). Only seven settlements, all located in Sakha (Yakutia), did not experience climate warming between the two periods.

Chukotka was the administrative region with the most pronounced warming, with over half of its area experiencing temperature increases

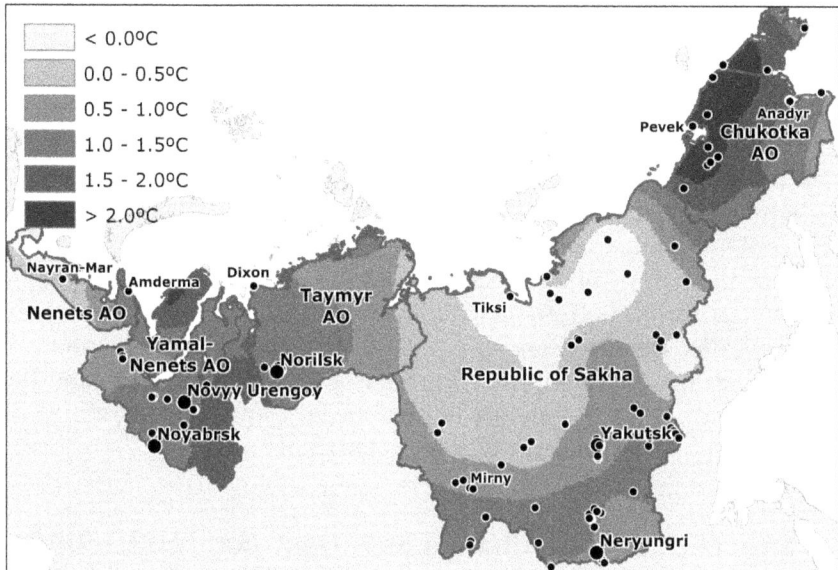

Map 9.2. | *Changes in Mean Annual Air Temperature between the 1970s and 2000s*

above 1.5°C. The northern part of Chukotka experienced the most significant increases, with temperatures rising more than 2°C. Of the 10 settlements to experience the greatest change between the two time periods, nine were within northern Chukotka. The highest change occurred in the settlement of Leningradskiy, a gold mining town of 760 people in 2002 (no permanent population remained by 2010), where the temperature increased by 4°C. The majority of the settlements experienced temperature increases above 2°C, including Pevek, Mys Schmidta, Bilibino, and Egvekinot. Cherskiy (Sakha), which is relatively close to these settlements (directly across Chukotka's western border in Sakha), experienced a warming of 2°C. Together, these settlements accounted for a population of 27,000 in 2002. Other areas with considerable warming include parts of the Yamal Peninsula in the northern Yamal-Nenets AO, with increases above 1.5°C. Large population centers present particular interest. Of the biggest cities, Norilsk has experienced the most significant change in temperature, 1.4°C. Mean annual air temperature in Novyy Urengoy and Noyabrsk increased by 1.3°C. Neryungri and Yakutsk experienced 0.9°C and 0.8°C increase respectively.

While the primary trend has been a pronounced warming, some areas also exhibited no change or even a decrease in air temperatures.

Across the study area, temperatures decreased above the Arctic Circle in the western Nenets AO and northern Sakha, along with a small area in eastern Sakha. Only six settlements exhibited temperature decreases, all located in northern Sakha. The settlement with the most pronounced decrease between the 1970 and 2000 time periods was Tiksi with a decrease of almost 0.5°C. The remaining settlements were Severniy, Ust-Kuyga, Chokurdakh, Belaya Gora, and Deputatskiy. All six settlements had small populations, the largest of which is Tiksi, with a population of above 5,000. Together, these six settlements accounted for 18,000 people. Conversely, 79 settlements witnessed temperature increases above 0.5°C accounting for a population of 1.2 million people, demonstrating that for most settlements and the vast majority of the population of the Russian Far North, warming has been the primary climatic trend in recent history.

Effects of Climate Change on Infrastructure Stability

Though air temperatures are the main driver of change, warming temperatures alone do not translate uniformly to impacts on infrastructure and living conditions, as other bioclimatic indicators may intensify or offset the impacts of these changes. However, temperature change sets an important baseline for quantitative impact assessment across the region. Effective permafrost engineering designs aim to protect permafrost from warming through isolation during the summer and by providing heat exchange between the permafrost and atmosphere during the winter. With adequate maintenance, this exchange allows for relatively stable permafrost temperatures (while even occasionally resulting in temperature decreases). In the ideal case, the permafrost temperature beneath structures built using the passive method with an elevated first floor will approach the mean annual air temperature of the surrounding area.

Actual conditions, however, differ significantly from the ideal. A substantial number of reports on the deformation of structures built on permafrost, both in the scientific literature (Kronik 2001, ACIA 2005) and the public media, have prompted speculation that climate warming is at least partially responsible. While this chapter focuses on a broad geography, as site-specific studies are largely beyond the scope of the current work, it is important to note that the typical Soviet-era five story apartment building has on average 80 households totaling about 185 residents. Structural problems with only a few of these buildings can become a serious issue as hundreds of residents have to be relocated. In many cases, the urban indigenous population has the fewest opportunities and smallest amount of economic resources to relocate.

Our results indicate that observed climate warming has the potential to decrease the bearing capacity of permafrost foundations built in the 1970s in the majority of settlements across all five study regions. Substantial decreases in potential foundation bearing capacity occurred in the regions of eastern Chukotka, southern parts of the Yamal-Nenets AO, and the Sakha Republic (Map 9.3).

The largest decreases in bearing capacity occurred in eastern Chukotka along the Pacific Ocean, southern Yamal-Nenets AO, and the southeastern tip of the Sakha Republic. In these areas, settlements saw a catastrophic drop in bearing capacity of more than 20 percent from the 1970s to the 2000s period. Settlements including Noyabrsk, Muravlenkovo, and Tarko-Sale in the Yamal-Nenets AO; Beringovskiy and Provideniya in Chukotka; and Vitim and Peleduy in Sakha were affected, totaling 166,000 people in 2002. Of these areas, most concerning are the three settlements in the Yamal-Nenets AO, which accounted for a population of 151,000. Loss of infrastructure stability in this region is particularly problematic, as the majority of the Russian oil and gas infrastructure is located in the Yamal-Nenets AO, as are the largest energy-producing urban settlements.

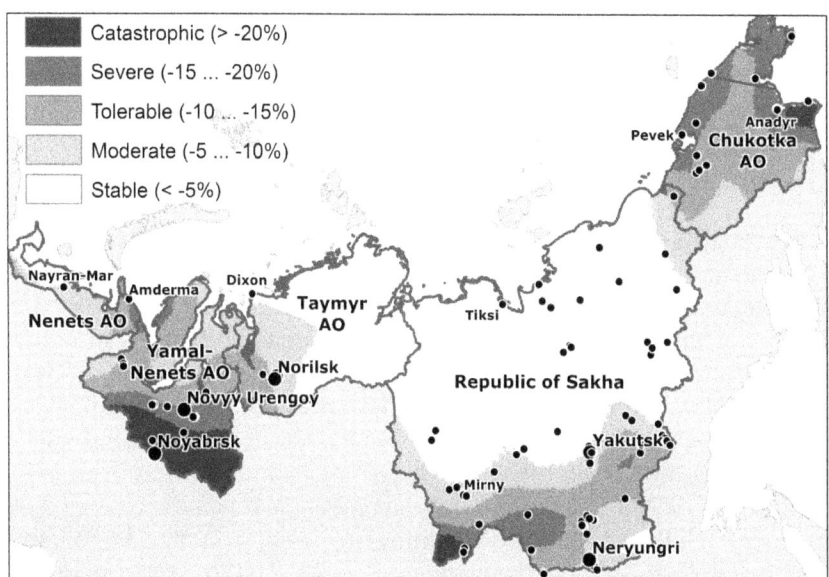

Map 9.3. | *Changes in Foundation Bearing Capacity between the 1970s and 2000s*

Areas least affected were 22 settlements considered stable, with a bearing capacity percent change of less than 5 percent. All except one (Dikson, in Taymyr) were located in northern or central Sakha. Of the four settlements that experienced increases in bearing capacity relative to the 1970 period, all were located in northern Sakha, including Severniy, Belaya Gora, Deputatskiy, and Ust-Kuyga.

The biggest cities with the largest decreases were Noyabrsk (28 percent) and Novyy Urengoy (15 percent). Norilsk, Neryungri, and Yakutsk all experienced moderate changes, decreasing 10, 9, and 6 percent, respectively. According to our estimates, climate-induced decreases of bearing capacity of 15–20 percent occurred in Salekhard and Nadym, and smaller decreases of 5–10 percent occurred in Lensk and Cherskiy. Anadyr and Pevek experienced the most severe decreases, with more than 20 percent loss of bearing capacity.

While climate warming is a plausible cause for decreases in foundation bearing capacity, other technogenic factors, such as inadequate structural design or lack of proper maintenance, should also be considered. Undetected leaks in sewage and water pipes are well known to result in rapid warming and chemical contamination of permafrost below building foundations. The resulting decrease in the soil's ability to support foundations has resulted in serious deformation of many structures (Grebenets, Streletskiy, and Shiklomanov 2012). While the in-depth investigations that would be required to assess foundation deformation caused by technogenic factors are well beyond the scope of this study, our results clearly indicate that climate warming has a significant potential for undermining the structural stability of structures built on permafrost.

Role of Nonclimatic Factors on Infrastructure Stability

A wide variety of foundation impacts are related to environmental effects of the built environment, most noticeably changes in conditions at the ground surface, such as those related to snow and vegetation. All settlements experience changes in radiation balance due to changes in their surface albedo, which in turn affect wind speed and direction, playing a role in snow redistribution. Moreover, even relatively small Arctic settlements can act as heat islands (Klene, Nelson, and Hinkel 2012; Konstantinov, Grishchenko, and Varentsov, 2015). Another problem is changes in snow albedo due to pollution in industrial centers of the Arctic. In the cities of Vorkuta and Norilsk snowmelt occurs a month earlier than in surrounding areas due to the accumulation of dust particles from coal and metallurgy plants. Soil salinization and waterlogging is another problem facing some of the Russian Arctic cities built on permafrost, particularly those with developed mining and metallurgy industries; Norilsk and Vorkuta again fall into this category. Technogenic salinization is not only leading to decreases in the stability of infrastructure through increases in active layer thickness and temperature rises in soil, but also directly affects foundations through the corrosion of metal and concrete in the active layer. Another major problem is excess heat from numerous underground utility pipes, which slice frozen ground into relatively small strips, isolating patches that warm faster than otherwise larger frozen areas would (Grebenets, Streletskiy, and Shiklomanov 2012). In this section, we briefly outline the importance of changes in snow cover and vegetation conditions in urban areas. Readers interested in a detailed explanation of the diverse impacts of permafrost degradation can find them in related publications (Anisimov et al. 2010; Khrustalev, Parmuzin, and Emelyanova 2011; Shiklomanov and Nelson 2013; Streletskiy, Anisimov, and Vasiliev 2014).

In order to maintain accessibility to key urban services, such as roads, driveways, and parking lots, other areas are used as snow dumps—yards, parks, rivers, and lakes. Not surprisingly, a redistribution of snow in populated areas leads to changes in ground temperatures and may intensify some of the ongoing cryogenic processes. As a rule, areas with higher snow accumulation will have warmer ground temperatures, as more snow provides more thermal insulation in winter and areas with no snow will have colder ground temperatures. Consistently warmer ground temperatures and excessive water resulting from snowmelt may lead to permafrost degradation, melting of ground ice, associated ground subsidence, and thermokarst development. Consistently colder temperatures intensify cryogenic weathering, frost cracking, and frost heave.

For example, removal of snow from roads leads to an increase of cold flux in the ground during the cold period of the year, resulting in lower ground temperatures relative to natural settings. Observations in Norilsk have shown that in the areas used as snow dumps, the permafrost temperature is 2–3°C higher than under consistently cleared roads (Grebenets, Streletskiy, and Shiklomanov 2012). Colder permafrost temperatures along roadways alter surface runoff, since road pads act as frozen dams, leading to waterlogging along the roads during the warm season. While areas along the roads accumulate standing water in the summer, leading to thermokarst development in the long term, the roads themselves experience an intensified process of frost cracking, differential frost heave, and subsidence, resulting in uneven road surfaces. These processes significantly limit the life span of the roads in the Arctic, requiring expensive maintenance and creating potentially dangerous driving conditions. Unfortunately, there are no effective extant methods to mitigate waterlogging along Arctic roads, so the monitoring of road beds is essential.

Similarly to snow, vegetation provides thermal insulation to the ground. Despite the fact that vegetation provides less ground insulation than snow, the influence of vegetation on the ground temperatures is complex, especially in high shrub and forest environments. Not only does it act as a thermal insulator between the atmosphere and the ground, but it can also play a substantial role in the redistribution of snow cover (Kudryavtsev et al. 1974). Snowmelt under forest canopies is different than in open environments, as the overlying canopy intercepts radiation and suppresses wind. As a result, melt rates are lower in forests than in equivalent open areas. Higher snow accumulation was found in areas of low shrubs, compared to those with tall shrubs and sites without vegetation. An important role in heat exchange at the surface in Arctic environments is attributed to moss cover as the presence of moss leads to a lower mean annual temperature protecting permafrost from thawing (Walker et al. 2003). Moss cover has been demonstrated to significantly reduce ground temperatures and thaw depth, demonstrating its importance to the overall ground temperature regime (Kade and Walker 2008).

The destruction of vegetation and moss cover in urban areas increases heat flow into the ground, leading to increases in active layer thickness and permafrost temperatures, thereby decreasing the stability of the structures built on permafrost. The removal of vegetation further intensifies erosion processes on slopes. Changes in vegetation extent can exceed the development area by several times, for example, dead taiga near Norilsk extends tens of kilometers from the city limits due to extensive pollution spreading from urban factories. The vegetation disturbance was found to exceed areas of gas production in West Siberia by 3–5 times (Garagulya et al. 1996). Technogenic waterlogging also has resulted in the deterioration of biodiversity in affected forests of the taiga zone.

Conclusions and Perspectives

The Arctic urban population is at the frontier of climate change. Climate change is not uniform, but can be exacerbated in areas of concentrated human activities. The combination of technogenic impacts and warming climate can have severe socioeconomic consequences, especially in large population centers located on permafrost. Numerous studies documented that permafrost stores large amounts of carbon, however the impact of permafrost thaw on global climate system is relatively small, (Streletskiy, Anisimov, and Vasiliev 2014). At the same time, the impacts of climate and human-induced change on permafrost are a considerable threat to human development rather than a long-term concern and are largely neglected. One of the immediate impacts is the decrease in the ability of foundations built on permafrost to support buildings and structures, affecting a substantial portion of the population residing in the Arctic cities. Intensification of such processes as coastal erosion due to a longer ice-free period, waterlogging, and thermokarst development is likely to reshape tundra landscapes, with negative effects on accessibility and the functioning of northern communities. Therefore, a changing climate in the northern regions could profoundly affect the entire Russian economy as it depends heavily on the extraction and transportation of mineral resources from the Arctic. More detailed interdisciplinary studies, including integration of traditional knowledge, establishment of long-term monitoring of permafrost in urban areas, and comprehensive evaluation of the relative role of climatic and human-induced impacts on infrastructure on permafrost are required in order to develop sustainable strategies for Russia's Arctic cities.

Dmitry Streletskiy is Assistant Professor of Geography and International Affairs at the George Washington University.

Nikolay Shiklomanov is Associate Professor of Geography and International Affairs at the George Washington University.

Notes

This research is supported by the Russian Science Foundation, project 14-17-00037, NSF grants PLR- 1231294, ICER-1558389, and OISE-1545913.

References

ACIA. 2005. *Arctic Climate Impact Assessment*. Cambridge: Cambridge University Press.

Anisimov, O.A., M.A Belolutskaya, M.N. Grigoriev, A. Instanes, V.A. Kokorev, N.G. Oberman, S.A. Reneva, Y.G. Strelchenko, D.A. Streletskiy, and N.I. Shiklomanov. 2010. *Major Natural and Social-Economic Consequences of Climate Change in the Permafrost Region: Predictions Based on Observations and Modeling.* Moscow: Greenpeace.

Anisimov, O.A. and D.A. Streletskiy. 2015. "Geocryological Hazards of Thawing Permafrost," *Arktika XXI Century* 2(3): 60-74 (in Russian).

Bogdanov, N.S. 1912. "Permafrost and Structures Resting on It." Unpublished manuscript.

CNR. 1990. "Stoitelnie Normi i Pravila [Construction norms and regulations]. Foundations on Permafrost 2.02.04–88." Moscow: State Engineering Committee of the USSR.

Garagulya, L.S., G.I. Gordeeva, E.D. Ershov, and N.I. Trush. 1996. "Characteristics of Distribution of Hazardous Geocryological Processes in Russian Cryolithozone," *Vestnik MGU, Seriya Geologiya* 4: 94–103.

Grebenets, V.I., D.A. Streletskiy, and N.I. Shiklomanov. 2012. "Geotechnical Safety Issues in the Cities of Polar Regions," *Geography, Environment, Sustainability* 5(3): 104–19.

Kade, A. and D.A. Walker. 2008. "Experimental Alteration of Vegetation on Nonsorted Circles: Effects on Cryogenic Activity and Implications for Climate Change in the Arctic," *Arctic, Antarctic, and Alpine Research* 40(1): 96–103.

Kalnay, E., M. Kanamitsu, R. Kistler, W. Collins, D. Deaven, L. Gandin, M. Iredell, S. Saha, G. White, and J. Woollen. 1996. "The NCEP/NCAR 40-Year Reanalysis Project," *Bulletin of the American Meteorological Society* 77(3): 437–71.

Khrustalev, L.N. and I.V. Davidova. 2007. "Forecast of Climate Warming and Account of It at Estimation of Foundation Reliability for Buildings in Permafrost Zone," *Earth Cryosphere* 11(2): 68–75.

Khrustalev, L.N., S.Y. Parmuzin, and L.V. Emelyanova. 2011. *Reliability of Northern Infrastructure in Conditions of Changing Climate.* Moscow: University Book Press.

Klene, A.E., F.E. Nelson, and K.M. Hinkel. 2012. "Urban–Rural Contrasts in Summer Soil-Surface Temperature and Active-Layer Thickness, Barrow, Alaska, USA," *Polar Geography* 36(3): 1–19.

Kondratiev, V.G. 2013. "Geocryological Problems of Railroads on Permafrost," *ISCORD 2013: Planning for Sustainable Cold Regions: Proceedings of the 10th International Symposium on Cold Regions Development, June 2–5, 2013.* Anchorage, AK: ASCE Publications.

Konstantinov, P.I., M.Y. Grishchenko, and M.I. Varentsov. 2015. "Mapping Urban Heat Islands of Arctic Cities Using Combined Data on Field Measurements and Satellite Images based on the Example of the City of Apatity (Murmansk Oblast)," *Atmospheric and Oceanic Physics* 51(9): 992–998.

Kronic, Y.A. 2001. "Accident Rate and Safety of Natural-Technogenic Systems in Cryolithozone." *Proceedings of the Second Conference of Geocryologists of Russia, Moscow.*

Kudryavtsev, V.A., L.S. Garagula, K.A. Kondrat'yeva, and V.G. Melamed. 1974. "Osnovy merzlotnogo prognoza." Moscow: Izdatel'stvo MGU.

Oberman, N.G. 2008. "Contemporary Permafrost Degradation of the European North of Russia," in D.L. Kane and K.M. Hinkel (eds), *Proceedings of the Ninth International Conference on Permafrost. Institute of Northern Engineering, University of Alaska Fairbanks, 29 June–3 July, Fairbanks, AK.*

Romanovsky V.E., S.L. Smith, H.H. Christiansen, N.I. Shiklomanov, D.A. Streletskiy, D.S. Drozdov, G.V. Malkova, N.G. Oberman, A.L. Kholodov, and S.S. Marchenko. 2015. "Terrestrial Permafrost [in 'State of the Climate in 2014']," *Bulletin of the American Meteorological Society* 96 (7): S139-S141
Romanovsky, V.E., D.S. Drozdov, N.G. Oberman, G.V. Malkova, A.L. Kholodov, S.S. Marchenko, N.G. Moskalenko, D.O. Sergeev, N.G. Ukraintseva, and A.A. Abramov. 2010. "Thermal State of Permafrost in Russia," *Permafrost and Periglacial Processes* 21(2): 136–55.
Rosstat. 2014. Census of Russia. Retrieved from http://www.gks.ru.
Shiklomanov, N.I. 2005. "From Exploration to Systematic Investigation: Development of Geocryology in 19th- and Early-20th-Century Russia," *Physical Geography* 26(4): 249–63.
Shiklomanov, N.I. and F.E. Nelson. 2013. "8.22 Thermokarst and Civil Infrastructure," in John F. Shroder (ed.), *Treatise on Geomorphology*. San Diego, CA: Academic Press.
Shiklomanov, N.I., D.A. Streletskiy, and F.E. Nelson. 2012. "Northern Hemisphere Component of the Global Circumpolar Active Layer Monitoring (CALM) Program," in *Proceedings of the 10th International Conference on Permafrost*. Salekhard, Russia.
Shur, Y.L. and D.J. Goering. 2009. "Climate Change and Foundations of Buildings in Permafrost Regions," in Rosa Margesin (ed.), *Permafrost Soils*. New York: Springer.
Stephenson, S.R., L.C. Smith, and J.A. Agnew. 2011. "Divergent Long-Term Trajectories of Human Access to the Arctic," *Nature Climate Change* 1(3): 156–60.
Streletskiy, D., N. Shiklomanov, V. Kokarev, and O. Anisimov. 2015. "Arctic Cities, Permafrost and Changing Climatic Conditions," in *Proceedings of the 68th Canadian Geotechnical Conference - GEOQuébec 2015*. Québec, Canada. September 20–23. Paper 123.
Streletskiy, D.A., O.A. Anisimov, and A.A. Vasiliev. 2014. "Permafrost Degradation," in Wilfried Haeberli and Colin Whiteman (eds), *Snow and Ice-Related Hazards, Risks, and Disaster*. Amsterdam: Elsevier, pp. 303–44.
Streletskiy, D.A., N.I. Shiklomanov, and E. Hateberg. 2012. "Infrastructure and a Changing Climate in the Russian Arctic: A Geographic Impact Assessment," *Proceedings of the 10th International Conference on Permafrost* 1: 407–12.
Streletskiy, D.A., N.I. Shiklomanov, and F.E. Nelson. 2012a. "Permafrost, Infrastructure, and Climate Change: A GIS-Based Landscape Approach to Geotechnical Modeling," *Arctic, Antarctic, and Alpine Research* 44(3): 368–80.
———. 2012b. "Spatial Variability of Permafrost Active-Layer Thickness under Contemporary and Projected Climate in Northern Alaska," *Polar Geography* 35(2): 95–116.
Vasiliev, A.A., M.O. Leibman, and N.G. Moskalenko. 2008. "Active Layer Monitoring in West Siberia under the CALM II Program," in L. Kane and K.M. Hinkel (eds), *Proceedings of the Ninth International Conference on Permafrost*. Institute of Northern Engineering, University of Alaska Fairbanks, 29 June–3 July, Fairbanks, AK.
Walker, D.A., G.J. Jia, H.E. Epstein, M.K. Raynolds, F.S. Chapin III, C. Copass, L.D. Hinzman, J.A. Knudson, H.A. Maier, and G.J. Michaelson. 2003. "Vegetation-

Soil-Thaw-Depth Relationships Along a Low-Arctic Bioclimate Gradient, Alaska: Synthesis of Information from the ATLAS Studies," *Permafrost and Periglacial Processes* 14(2): 103–23.

Washburn, A.L. 1980. *Geocryology: A Survey of Periglacial Processes and Environment.* New York: Wiley.

Whiteman, G., C. Hope, and P. Wadhams. 2013. "Climate Science: Vast Costs of Arctic Change," *Nature* 499 (7459): 401–3.

Zhang, T., R.G. Barry, K. Knowles, J.A. Heginbottom, and J. Brown. 2008. "Statistics and Characteristics of Permafrost and Ground-Ice Distribution in the Northern Hemisphere," *Polar Geography* 31(1–2): 47–68.

CHAPTER TEN

Urban Vulnerability to Climate Change in the Russian Arctic

Jessica K. Graybill

Introduction

As the global community learns that the Arctic is one of the most rapidly transforming regions due to global climate change and that cities are some of the most vulnerable places to climate transformations, an investigation of climate concerns for Arctic cities becomes pertinent to understanding how human communities, the built environment, and ecosystems might fare into the future with ongoing climate change and climate-related transformations.[1] The Russian Federation contains the most Arctic territory, the most highly urbanized places in the Arctic, and greatest number of urban residents across all of the Arctic. Unraveling the meaning of current and impending climate transformations for the Russian Arctic, and urban places, requires an understanding of interrelated human and natural systems and concerns related to climate change that are biophysical, socioeconomic, political, and cultural. Because Russia's Arctic urban environments are complex socio-environmental systems that are simultaneously human and biophysical, this type of environment is one of the most dynamic and difficult socio-ecological systems to understand, maintain, and predict. Understanding the specific concerns regarding climate change for Russian Arctic cities requires simultaneous attention to multiple complex systems.

Worldwide, urban places have begun to address climate change concerns for the present and future. Often, these tasks involve multiple stakeholders and suggest long-term engagement with climate futures for people in specific urban locales (Sclar et. al. 2013; Baker 2012; Hoornweg et al. 2011; Rosenzweig et al. 2011). One increasingly common type of evaluation of risks in cities due to climate change is an urban vulnerability assessment (UVA) (Rosenzweig et al. 2011). Assessing vulnerability attempts to measure the degree to which environments or communities are susceptible to, or unable to cope with, the adverse impacts of climate change. The degree of vulnerability determines how severe the impacts of climate change might be. Thus, attention to climate change vulnerability in urban settings has become an increasingly important topic; globally, the urban population is currently over 50 percent and is even higher within the Russian North (Arctic and sub-Arctic places). The goal of a UVA is to induce policy making locally and regionally that directly addresses climate concerns in urban places for people, the built environment, and nature in the city.

In this chapter, I explore the social and political milieu for addressing climate change in Russia to understand how local vulnerabilities to climate change are produced for urban settings in the Russian North. I address concerns of both Arctic and sub-Arctic places because the sociopolitical and economic challenges in these two proximal regions of Russia have similar past development trajectories, face similar concerns into the future regarding climate change, and have "lessons" applicable to both regions. First, I provide a brief overview of Russia's climate leadership internationally and domestically with a focus, where possible, on the North. This overview provides a sketch of the kinds of attention to climate change that might be expected among Russian citizens, scientists, and the government. Second, I address perceptions of climate change in the Russian North to illustrate how people in this region consider their natural, built, and social environments. Addressing local and indigenous peoples' perceptions of climate change provides insight into how or why policy making addressing these issues is (or is not) occurring locally. Third, I present a conceptual climate vulnerability framework that could be applied to Russian Arctic cities to elucidate the range of exposures, sensitivities, and potential adaptive capacities to be addressed in Arctic urban climate policy. The climate vulnerability model presented helps indicate the degree and kinds of vulnerabilities—and resiliencies—that could be addressed in urban climate policy in Russian Arctic cities.

Russia's Climate Leadership

Climate science has a rich history and future worldwide, but especially in Russia. Russia long has been a leader in the development of climate science. Throughout the territorial expansions and contractions across Eurasia of the Russian Empire, the Soviet Empire, and now the Russian Federation, Soviet and Russian trained scientists have contributed to over three hundred years of climate knowledge about this world region through field and laboratory science endeavors. Prior to the Soviet century, explorers and ethnographers described Eurasian places and people regarding the impacts of weather and climate on daily life, subsistence survival, and possibilities for the expansion of empire (exploration of Kamchatka: Krasheninnikov 1972; exploration of Siberia: Arseniev 1996). Soviet science produced an army of well-trained observers and analysts of biological and physical conditions in varying climate regimes using multiple methods. Some examples include ecological modeling and dendrochronology of the extensive Ural mountain region (Shiyatov 1988), regional human and cultural adaptations to weather and climate on Kamchatka (Sharkun 2008), and atmospheric climatologists who attempted to alter urban climates for special events (e.g., for the 1980 Moscow Olympics, see Oldfield 2013). Current climate science in formerly Soviet Eurasia continues in prior traditions, including climate geoengineering, but largely resists embracing some Western climate science knowledge, especially that which assumes anthropogenic (human-induced) climate change as a starting point for explaining biogeophysical and social transformations in Eurasia today. This chapter explores some of the background to and reasons for this resistance by noting important aspects of Russia's engagement with climate science and climate knowledge in international and domestic (national and regional Arctic) arenas.

International Involvement in Climate Policy

Widespread publicity by international scientists and activists of the extent and pace of climatic changes worldwide has increased greatly since the late 1980s. An early and public discussion of the transformations due to anthropogenic climate change that could affect our planet occurred in 1988, when NASA climatologist James Hansen testified, in US congressional hearings, to (1) the sweeping changes that climate change would make to natural and human systems and (2) the importance of recognizing human-induced (anthropogenic) climate change as a driver of change for these systems. Since the 1980s, and with increasing attention to climate change concerns internationally, climate science has become an important yet also controversial and seemingly all-encompassing concern for scientists and governing bodies at all scales (national, regional, local) who seek to understand the magnitude and pace of human and environmental transformations occurring due to climate change. Today, addressing this issue lies at the heart of policymaking for many national, regional, and local entities because governments recognize climate as an important driver of stability—or change—of environmental and human systems, especially in sensitive regions such as the Arctic (IPCC 2014).

For Russia, participation in international climate policy talks and developing national policy during the late twentieth and into the early twenty-first century has had a unique path. Socioeconomically, while most international participants in climate policy development could easily be categorized as representing "developed" or "developing" places, Russia's path and thus participation was shaped by the collapse of the Soviet Union in 1991. In the 1990s, the transitional economic status of this region was somewhere in between developed and developing, as the new Russian Federation and the neighboring Newly Independent States struggled to regain footing amidst socioeconomic, political, and cultural transformation. Politically, in terms of international policy making, Russia's years of economic crisis in the 1990s meant that the previously heavily industrialized and "developed" Soviet Union had decreased its manufacturing activities, creating a country profile that looked much more like a "developing" country in terms of its carbon footprint (Lioubimtseva 2010). Reductions in greenhouse gas emissions between 1990 and 1998, for example, went from 3.32 billion metric tons to 1.98 billion metric tons (IEA 2010).

Perhaps Russia's most memorable moment in international climate talks was during the ratification of the Kyoto Protocol. While Russia signed the protocol in 1999, ostensibly agreeing to the importance of addressing climate change through greenhouse reductions, the majority of climate-related policy makers were against any restrictions in industrial activity that would harm the Russian economy. Indeed, during the World Climate Change Conference in Moscow, 2003, Russia's four-decade leader of international environmental fora, Yuri Izreal, decidedly avoided the topic of ratification of the protocol, leading to the press labeling him as "the communist fossil fighting for fossil fuel" (Josephson et al. 2013). Early in January 2013, Russia also made the decision to discontinue participation in the protocol, noting that the other major producers of greenhouse gases (United States, China,

India) were also not committed to reductions. Thus, Russia became empowered during the creation of the Kyoto Protocol by the generous allotment of emissions (based on Soviet-era industrial activity) and accumulated carbon credits (based largely on 1990s-era transition economic status). This unique positioning on the international scene has given Russia great power over the future climate policy architecture and, internationally, the "image as a 'savior' of the Protocol and to indicate Russia's unity with 'European politics' or 'Western values'" (Turkowski 2012: 4).

Currently, accumulated carbon credits and increasing emissions fuel Russia's industrial activity without the nation being crushed by international legislation that could hinder national economic progress. Even after the 2015 United Nations Climate Change Conference in Paris (COP21), Russia's proposed "commitment" to greenhouse gas emissions reduction "allows greenhouse gas emissions excluding [forested land cover] to grow significantly from the current levels" (Climate Action Tracker 2015). Most important to notice, after COP21, are three pivotal factors that shape Russia's climate policy: (1) there is uncertainty about how emissions from land use are accounted for in the current emissions inventory produced by Russia; (2) energy production remains a central focus of national policy; and (3) while different programs and policies exist on paper to achieve greater energy efficiency, "implementation is lagging" (Climate Action Tracker 2015: 3). In fact, experts from the World Resources Institute suggest that Russia "could increase its emissions about 40–50 percent above current levels by 2030" (World Resources Institute 2015) due to the framing of Russia's Intended Nationally Determined Contribution (INDC) target as a reduction from 1990 levels and not from actual levels produced in the post-Soviet period.

Differently from most of the other countries of the former Soviet Union, Russia has always been in compliance by reporting its obligations and submitting its national communications on climate change on, or ahead of, time to the United Nations Framework Convention on Climate Change (in 1995, 1998, 2002, 2006, and 2014). Even during the most difficult period of socioeconomic and political transformation after the collapse of the Soviet Union, Russia perceived its involvement in the international climate negotiations as a source of national pride and an important component of the ongoing status of being a superpower (Henry and Sundstrom 2007; Lioubimtseva 2010). Scientifically, the Soviet Union's and Russia's long histories of a well-developed program of climate science across Eurasia is respectable and deserves greater in-depth inspection for understanding changes in climatic trends across Eurasia over the last several centuries. But, the rise of the need for inspection came at the same moment as the disintegration of the Soviet Union and the creation of multiple nation-states across Eurasia. Many of the peripheral regions of the former Soviet Union have had varying ability to continue investing in existing climate science programs (indeed, in all of science; Graham and Dezhina 2008; Graham 1993) or have been unable or uninterested to invest—socially or economically—in the creation of nationally based programs (Sabonis-Helf 2002). Climate change science about the Russian Arctic only recently became important to today's Russian government (Putin 2010), and this region once again provides a stage for Russia to showcase its prominence in the world of climate science. In the realm of Arctic science, Russia is making a comeback as an active player in the international science and policy spheres through increased scientific explorations based beyond Russia's borders and by increased activity in pan-Arctic agenda setting through entities such as the Arctic Council and the International Arctic Science Committee, among others.

Domestic Acknowledgement of Climate Change

Despite its ostensible successes in the international sphere, Russia's domestic climate policy for the last several decades has been poorly defined. Several governing bodies that have been tasked with coordinating climate concerns among different branches of government nationally have had varying rates of success. Some continue to exist and have grown, while others have shrunk or simply been eliminated. For example, the Ministry of Natural Resources has become larger and more powerful through political reorganization during the 2000s, such as the dissolution of the State Committee on the Protection of Nature in 2000 (Josephson et al. 2013), which was then reconstituted as the State Committee on Environment (Goskompriroda) within the Ministry of Natural Resources and Ecology. Now *Goskompriroda*, which has no ability to make rules or regulations, functions mostly as an environmental advocacy committee. Furthermore, local-level governance remains dependent on federal funding and guidance, a remnant of the Soviet system. Due to the constant flux of federal-level rules and regulations and lengthy and multiple federal-level reorganizations of scientific and political institutions since 1991, policy making and agenda setting at regional and local levels have been hampered. In many cases, progress towards addressing environmental challenges—including climate change—has stagnated (Josephson et al. 2013).

Additionally, internal battles within government agencies over leadership of climate knowledge building have slowed efforts to develop climate policy (Sabonis-Helf 2002). Since 1994, the Interagency Commission has existed to coordinate efforts. It is headed by the Russian Hydrometeorological Service (RosHydromet), which is a science- and monitoring-oriented government agency. Because of its focus on observations rather than projections and on technological know-how instead of policy development for Russia's regions, leadership by this agency on climate-related issues has concerned many researchers who worry about the seriousness of its efforts to understand how climate change links to energy, economic, and international trade realms (Sabonis-Helf 2002).

Some leadership regarding climate change has come from two sectors outside the government: corporations and nongovernmental organizations (NGOs). The private sector is not involved, directly, in creating climate policy, but with greater numbers of joint venture projects with European and Asian corporations, the Russian private sector has increasingly recognized the need to address climate change in economic (industrial) decision and policy making (Korppoo 2008; Turkowski 2012). Some corporate leadership regarding climate change includes voluntary tracking of greenhouse gas emissions and reductions within the industrial sector. Increasing opportunities for carbon trading with European and Asian partners have recently caused some important Russian corporations to become more involved in domestic discussions of climate change and control of it. For example, Center Telecom, the Federal Grid Company of Unified Energy System, Gazprom, Irkutskenergo, Novatek, and Tatneft were Russian corporations leading efforts to understand and tackle climate change concerns, according to the 2009 Carbon Disclosure Project for Russia (Carbon Disclosure Project 2010).

Additionally, nongovernmental organizations, both Russian and international, are involved in trying to increase interest in creating domestic climate policy or interest in the topic (specifically, see world.350.org/russian). But many NGOs can play only minor roles in advocating for change in Russia today because of the political constraints placed on them. NGOs across Russia face increased scrutiny since new legislation introduced in 2012 makes it possible to designate civil society actors who criticize governmental measures as "foreign agents" and individuals involved in such organizations can be legally prosecuted as traitors or terrorists (The Russia Monitor 2012). Such punitive measures, combined with a domestic climate policy arena that is limited in scope and dominated by few—and entrenched (Sabonis-Helf 2002)—stakeholders means that the realm of forward-looking climate policy is slow to develop. Overall, it is widely noted that the nongovernmental sector has a limited impact on the direction of Russian climate policy (Sabonis-Helf 2002; Henry and Sundstrom 2007, Andonova and Alexieva 2012).

Domestic participation in internationally led climate groups is also minimal. Two urban-focused groups that have worldwide presence are the World Council of Mayors and Cities Climate Leadership Group (C40). The goal of C40 is to offer "cities an effective forum where they can collaborate, share knowledge and drive meaningful, measurable and sustainable action on climate change" (www.c40.org). Within Russia, only Moscow has become part of C40, which makes some sense, as this organization is mostly comprised of world capitals. Given the size of the Russian Federation, however, more active involvement from other Russian cities would be important. A more powerful and active group for urban settings is the World Mayors Council on Climate Change (www.worldmayorscouncil.org), whose purpose is to create an "alliance of committed local government leaders concerned about climate change" who "advocate for enhanced engagement of local governments as governmental stakeholders in multilateral efforts addressing climate change and related issues of global sustainability." Many of the world's urban places are involved in these social and political networks that seek to address climate change at the urban scale, but that coverage within Russia, including the Russian Arctic, is missing.

Domestically, there are three main concerns regarding climate change. First, there is limited public awareness and interest in the topic, partially due to a long history of rejecting the idea that humans could be a significant driver of climate change (Rowe 2009; Lioubimtseva 2010) alongside the Soviet-era notion instilled in the population, as part of the idea of progress, that technology would be able to solve environmental problems in the future (Graybill 2013b; Oldfield 2013). Second, and due to lasting hierarchical management, there is limited participation of multiple stakeholders in policy formation. During the Soviet era, environmental planning and management directives came from within the centralized government (Pryde 1991) and rarely were local initiatives for understanding or bettering the environment undertaken, except by local interest groups (such as clubs; Henry 2010). Third, competition among agencies for control and funding of climate science initiatives (e.g., scientific studies, policy) has prevented a concerted effort to address this concern within the government. The Russian parliament has begun to assert a voice in climate change matters alongside competition among regional and business entities (Lioubimtseva 2010). Whether there really is a commitment to climate change mitigation within Russia's policymaking agendas nationally is questionable, but what remains indisputable is that the discussion continues to be "largely shaped by its national academic and business community's perception of the impacts of climate change on the Russian economy, its vulnerability and adaptations" (Lioubimtseva 2010).

Climate Change Awareness/Perceptions in the Russian North

This contextualization of the international and domestic arenas for Russia's engagement with climate science provides background for understanding how communities in the Russian Arctic are responding to the discourses and actualities of climate change. Russia's public, generally, does not regard anthropogenic climate change as an acute environmental problem (Lioubimtseva 2010). Pugliese and Ray (2009) find that 85 percent of Russians are aware of climate change (compared to 97 percent of Americans and 62 percent of Chinese), but only 39 percent perceive it as a serious personal threat (compared to 63 percent and 21 percent respectively in the United States and China).

Does skepticism or lack of interest in the topic of climate change also translate to the Russian North (Arctic and sub-Arctic regions)? A review of scholarly literature finds a few different answers. From the Arctic Yamal Peninsula to the inland Sakha Republic to the sub-Arctic Kamchatka Peninsula, there are a range of understandings and engagements with climate change issues. In Yamal, Forbes and Stammler (2009) find that Western concepts of climate change, wildlife management, or traditional ecological knowledge (TEK) do not have local buy-in, leading to conceptual mismatches on the ground among indigenous people who live in the region and scholars who attempt to understand environmental climate changes there. Instead, local and indigenous understandings of climate change in the Russian North require Western scholars working in this region to engage with different conceptualizations of environmental change. These authors warn that, unless Western scientists alter their conceptual frameworks and remain open to Russian scientific and local/indigenous management ideas, a common language for addressing socio-environmental changes in this region with local and indigenous peoples will be hard to find.

Among Vilui Sakha, in Yakutia, Crate (2008) also writes about different frameworks and ways of rationalizing changes to local environments and food sources due to what Western scientists call climate change. Attempts to increase resilience to environmental change, sometimes understood by this indigenous group as climate related, are present but follow a non-Western pathway for decision making, thereby also calling into question, locally, Western conceptualizations of the drivers and impacts of climate change.

On Kamchatka, Western-style engagement with the concept of climate change is nascent, and Sharakhmatova (2011) finds evidence of climate change effects on the local populations and environment through an examination of TEK in Itel'men communities in coastal western Kamchatka. In two mountain communities of local and indigenous peoples on Kamchatka, Esso and Anavgai, Graybill (2013b) found that interest in climate change fell far behind concern about other changes more prominent in everyday life, namely ongoing economic transition that continues to affect daily life situations, including transportation; critical food choices; and social networking. Indeed, some Koryak and Even indigenous peoples in this study found climate change to be an issue that was "all about politics," because changes in the climate system are "only natural" whereas discourse is "always about people" and their perspectives.

These case studies about climate change concerns from across the Russian North indicate that while knowledge about climatic changes exists among people who live in the Russian North, it is conceptualized differently than Western science might anticipate. While reception to the idea of climate change varied across all case studies, there is a common element: the dominant scholarly understandings of, and basis for, anthropogenic climate change is not widely accepted or understood by local and indigenous peoples living in this region. Combined with the lack of progress towards a strong domestic policy for acknowledging and addressing climate change within the Russian federal government, it seems that a policy-based agenda for the Russian North would be slow to develop unless Western ideas about climate change begin to engage local and indigenous understandings.

Climate Vulnerability in Russia's Arctic Cities

In a consideration of climate policy for the Russian North, it is necessary to consider not only the sparsely populated tundra and taiga but also the densely populated, compact urban settlements that dot the entire region. The discussion, above, of perceptions about climate change from western, central, and eastern locations across the Russian North provides even greater impetus to pursue this trajectory, as the people interviewed in each case study spend time in both urban and nonurban places while pursuing everyday and seasonal activities. Further contextualization is needed for understanding how urban places in this region may or may not be grappling with climate changes concerns. The fact that the population of the North is either largely unfamiliar or unconcerned with the rate and amount of climatic shifts occurring in this region in the recent (two to three decades) past, alongside a lack of domestic climate change–related leadership, leads to asking about the vulnerability of Russia's arctic urban places. How vulnerable will urbanized populations and settlements be to existing and projected climate changes? Providing a concrete answer to this question is beyond the scope of this chapter, but a conceptual framework aimed at addressing vulnerability is presented as a guide for examination of this topic.

Vulnerability studies have a long tradition in geographic and environmental sciences (Cutter et al. 2003; Adger 2006; De Sherbinin et al. 2007; McLaughlin and Dietz 2007). While the definitions of vulnerability are as numerous as vulnerability scholars, "vulnerability to climate change is the degree to which geophysical, biological and socio-economic systems are susceptible to, and unable to cope with, adverse impacts of climate change (IPCC 2014). Ultimately, the purpose of assessing vulnerability is to understand what is needed to build sustainability—through adaptive capacity—into a place or a phenomenon (Delica-Wilson and Willison 2004; Edwards 2010). While there are many ways to understand vulnerability, the IPCC (2014) framework is used here to express it, adding the urban component to further narrow this subject to our objects of concern, cities in the Russian Arctic:

Urban vulnerability = f (exposure, sensitivity, adaptive capacity)

In this conceptual formula, urban vulnerability is a function of three interrelated components. Exposure is considered to be climate stress, either climatic changes themselves or the object affected by the climate change. Sensitivity refers to the degree to which a system is affected by the climate stimuli. Adaptive capacity is the capability of the system—human and biogeophysical—to adjust to climate change. In this conceptualization of vulnerability, adaptive capacity acts as a mitigating force on the potential exposures and sensitivities that may exist in a given system (in this case, Russian Arctic cities). Some environments or communities may be highly vulnerable to low-impact climate changes because of high sensitivity or low adaptive capacity, while others can have little vulnerability to even high-impact climate changes because of insensitivity or high adaptive capacity. By understanding the specific concerns related to vulnerability within these three domains, decision and policy makers for urban places in the Russian Arctic could begin to understand the interplay among different aspects of vulnerability across the region, generally, and for individual urban locales, specifically.

A visualization of this conceptual model is provided in Figure 10.1. Each of the three domains of vulnerability (exposure, sensitivity, adaptive capacity) is represented in this diagram, and in each domain, major individual exposures, sensitivities, and potential adaptive capacities are represented. Peer-reviewed literature about Arctic climate hazards due to actual and projected biophysical changes inform the exposure section. The sensitivity section summarizes major con-

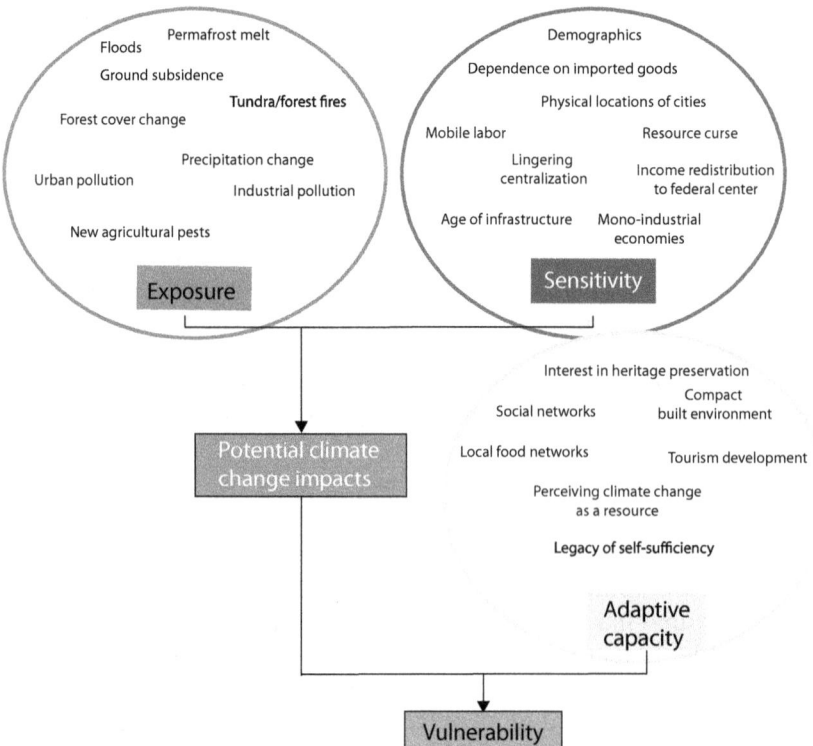

Figure 10.1. | *A Conceptual Model of Urban Vulnerability*

cerns in the literature related to concerns about the built and social environments of the Russian Arctic. In the adaptive capacity section, a range of literature about urban sustainability, urban ecology, anthropology, sociology, and geography about the Arctic and other regions inform the region's adaptive capacity. Because this diagram addresses human and biophysical components of vulnerability for cities in the Russian North, it may also serve as the basis for imagining the components of resiliency through adaptation in and for urban spaces in the Russian North. Such a conceptualization might become important to local and indigenous leaders, business, and policy makers concerned with managing or mitigating the range of exposures and sensitivities that are, or are projected to be, experienced in this region and its urban places. By adding adaptive capacities to the conceptualization, we might also understand how people in Russia's Arctic urban settlements can survive ongoing transformation to their biophysical and human environments by capitalizing on existing strengths (e.g., desire for heritage preservation) or networks (e.g., kin- and community-based food networks) instead of simply grappling with ongoing or impending changes due to existing sensitivities (e.g., permafrost change) and exposures (change to the built environment due to permafrost change and age of infrastructure).

Vulnerability and Urban Typology

In addition to exposure, sensitivity, and adaptive capacity, there is one additional aspect of Russia's Arctic cities that must also be considered regarding vulnerability. Pelyasov (2009) notes that it is imperative to understand the types of urban settlements found in the Russian North to understand the degrees of exposure, sensitivity, or adaptive capacity that each may experience. Types of urban settlements refers to multiple characteristics, including (but not limited to) settlement size and cultural and economic profile. There is a distinct typology of urban places in the Russian North that is easily distinguishable across the landscape. The Russian North was largely settled by Soviet-era newcomers in planned communities that were strictly regulated for population density and size (Morton and Stuart 1984; Graybill and Dixon 2012), and indigenous peoples were collectivized and placed in urban-type settlements during the Soviet era (Bartels and Bartels 1995). For example, Arkhangelsk is a large regional administrative center composed of universities, a diversified economy, and a population of ~350,000 people (2010 Census). Vorkuta, in contrast, is a monoindustrial town of ~115,000 people where salaries vary greatly between workers in the dominant, male-dominated, industry (coal mining) and service-type industries in the city (e.g., health services, shopkeeping). Perhaps the most vulnerable economies are those of small port cities, largely founded during the Soviet era of expansion into peripheral regions and where, in the post-Soviet era, federal subsidies and upkeep of port facilities have crumbled. These are places where out-migration is rampant and prospects for future development are low. Pevek, in Chukotka, is an example of such an urban settlement, which has experienced a threefold population decline since 1989, with ~4,000 residents as of 2010.

Last, urban-type settlements of mostly indigenous peoples cannot be forgotten in this typology. Often comprising only a few hundred residents, such as Anavgai on the Kamchatka Peninsula or Nogliki on Sakhalin Island, numerous indigenous urban settlements scattered across the Russian North live on as reminders of the era of collectivization and sedentarization enforced during the Soviet era. These urban-type settlements continue to be tolerated in the post-Soviet era by those now more familiar with the urban lifestyle, afforded by subsidized housing and access to physical (electricity, natural gas) and social (education, healthcare) resources, than with their traditional and/or indigenous ways. In such places, local economies are perhaps not as vulnerable as those of small, nonindigenous urban settlements because there are still relatively strong traditions of obtaining and maintaining local food through kin-based and social networks (Graybill 2013a, 2013b).

In each kind of urban setting, actual and potential adaptive capacity will vary because of the different structures, function, and historical trajectories of each locale. For example, cities like Arkhangelsk may become engines of regional innovation (e.g., Arctic creative cities; see Pelyasov 2009 and Petrov 2011) and provide new ways of thinking about adaptation through new technologies and ideas that serve the specific human and physical geographies of compact urban environments of the Russian North. Places like Vorkuta or Pevek, where economic transformation is occurring more slowly, may rely on social networks to reinvigorate slowly decentralizing management regimes with new local and regional talent that acknowledges the concerns of climate change. Even smaller settlements, such as Anavgai or Nogliki, may rely on existing kin-based regimes to assure food security for small, local communities experiencing changes to local environment-based food resources.

Concluding Thoughts

Despite ongoing concern about the Arctic as a region of rapid and sustained climate change, attention to cities in the Arctic region has been limited. While this has been the case across the Arctic region, it is especially true of the Russian Arctic. Long-term natural and social scientific studies indicate climatic changes in this region, yet there is limited public awareness and interest in the topic of climate change. Neglect regarding climate change has been both international and domestic and, while the situation is beginning to change in some larger cities in the Russian Arctic (see, for example, the strategic plan for Murmansk; UNDP 2009), there is almost no discussion regarding climate change as a basis for decision and policy making regionally, let alone for Russian Arctic cities. Governmental discussion of vulnerability and adaptation in the Arctic remains patchy at best (Loboda 2014). Potentially, this lack of attention creates dire circumstances for the hundreds of thousands of urban residents in Russia's Arctic, if climatic and associated biophysical, socioeconomic, and cultural changes continue at projected rates and geographic scales.

The lack of engagement with climate change in decision and policy making may be the result of two primary causes. First, it seems to be related to the lack of a domestic agenda for addressing climate change as an important driver of changes to human and environmental systems. The decline of science as an important national agenda item in Russia during the early 1990s occurred at roughly the same moment in history when climate change science took hold in the West as a major concern and funding initiatives to address it grew across disciplines and internationally (Moser 2010). Only when faced with new kinds of extreme weather aberrations, such as an abnormally cold Siberian winter in 2007 and the peat fires in southwestern Russia in 2010, did government leaders begin to consider anthropogenic climate change as a viable explanation of events (Tynkkunen 2010). Only very recently has funding for climate-related science in the Arctic regained momentum in Russia, with grants now being awarded for polar expeditions and climate observations by the Federal Service for Hydrometeorology and Climate Monitoring (Russian Geographical Society 2011).

Second, Rowe (2009) notes that Western and international climate science is considered in Russia as heavily influenced by political maneuvering and therefore of potentially reduced value because of policy, and thus political, involvement. Additionally, mistrust of the West, and thus of international policy, has persisted into the post-Soviet era. Due to national pride and distrust of Western scientific concepts, climate skepticism is important to address as part of the climate science landscape in post-Soviet Russia (Graybill 2013b). The resistance to "outsider" models in the Russian North occurs because Western and international conceptual models highlighting the importance of anthropogenic climate change do not match prior climate science understandings of climate change as regulated by natural cycles (Graybill 2013b) or existing local and indigenous knowledge of Arctic cultural and environmental transformations (Crate 2008; Forbes and Stammler 2009).

The lack of development of regional and local climate change policy in the Russian Arctic leaves urban settlements vulnerable to the effects of climate change. Specific elements of vulnerability for cities in this region are noted in the domains of exposure and sensitivity in Figure 10.1. Scientists, policy makers, and local communities need more work, conducted together, to fully comprehend all possible exposures and sensitivities for Russia's Arctic cities. There is a real need for a holistic, multiscalar approach to understanding the diversity of urban locales in this region before climate change–related policy can be developed. Creating such an approach would seem to involve at least four considerations. First, scholars and decision makers need a firm and up-to-date understanding of the overall urban typology for the Russian Arctic. Understanding the sizes, profiles (social and economic), and future prospects of individual urban locales is necessary to address vulnerability. Second, understanding how urban locales are interlinked with each other or with larger regions is imperative for policy formation. Third, successful development and lasting enforcement of policy is increasingly understood to be the work of multiple actors and stakeholders (Rozensweig 2011). In the case of the Russian North, involving existing governance structures, but also including corporate, nongovernmental actors, and local and indigenous leaders would seem appropriate for ensuring that the variety of concerns that might be experienced by regional peoples and places are addressed. Fourth, focusing attention on adaptive capacity alongside exposure and sensitivity is necessary to maintain vitality and a sense of hopefulness in this region undergoing immense change on multiple fronts (socioeconomic, political, cultural, and biogeophysical). Focusing on adaptive capacity provides scholars and community members ways to consider planning, preparing for, and implementing adaptation measures.

Jessica K. Graybill is an Associate Professor of Geography and Director of the Russian and Eurasian Studies Program at Colgate University.

Notes

1. An earlier version of this chapter appeared in *Polar Geography*. The author has updated this version extensively. Thank you to the editors of *Polar Geography* for permission to use the revised text.

References

Adger, W. Neil. 2006. "Vulnerability," *Global Environmental Change* 16: 268–81.
Andonova, Lilliana B., Alexieva, Assia. 2012. Continuity and Change in Russia's Climate Negotiations Position and Strategy. *Climate Policy* 12(5): 614-629.
Arseniev, V.K. 1996. *Dersu the Trapper.* Trans. M. Barr. New York: McPherson.
Baker, Judy L. (ed.). 2012. *Climate Change, Disaster Risk, and the Urban Poor: Cities Building Resilience for a Changing World.* Washington, DC: World Bank.
Bartels, D.A. and A.L. Bartels. 1995. *When the North Was Red: Aboriginal Education in Soviet Siberia.* Montreal and Buffalo: McGill-Queen's University Press.
Carbon Disclosure Project 2010. Retrieved 1 February 2016 from https://www.cdp.net/CDPResults/CDP-2010-G500.pdf.
Climate Action Tracker. 2015. Russian Federation. Retrieved 10 December 10 2015 from www.climateactiontracker.org/countries/russianfederation.html.
Crate, S. 2008. "Gone the Bull of Winter: Grappling with the Cultural Implications of and Anthropology's Role(s) in Global Climate Change," *Current Anthropology* 49(4): 569–95.
Cutter, Susan L., Bryan J. Boruff, and W. Lynn Shirley. 2003. "Social Vulnerability to Environmental Hazards," *Social Science Quarterly* 84(2): 242–61.
De Sherbinin, Alex, Andrew Schiller, and Alex Pulsipher. 2007. "The Vulnerability of Global Cities to Climate Hazards," *Environment & Urbanization* 19(1): 39–64.
Delica-Wilson, Zenaida and Robin Willison. 2004. "Vulnerability Reduction: A Task for the Vulnerable People Themselves," in Greg Bankoff, Georg Frerks, and Dorothea Hilhorst (eds), *Mapping Vulnerability: Disasters, Development and People.* London: Earthscan.
Edwards, Andres R. 2010. *Thriving Beyond Sustainability: Pathways to a Resilient Society.* Gabriloa Island, BC: New Society Publishers.
Forbes, B. and F. Stammler. 2009. "Arctic Climate Change Discourse: The Contrasting Politics of Research Agendas in the West and Russia," *Polar Research* 28: 28–42.
Graham, L. 1993. *Science in Russia and the Soviet Union: A Short History.* Cambridge: Cambridge University Press.
Graham, L. and I. Dezhina. 2008. *Science in the New Russia: Crisis, Aid, Reform.* Bloomington: Indiana University Press.
Graybill, Jessica K. 2013a. "Mapping an Emotional Topography of an Ecological Homeland: The Case of Sakhalin Island, Russia," *Emotion, Space and Society* 8: 39–50.
———. 2013b. "Imagining Resilience: Situating Perceptions and Emotions about Climate Change on Kamchatka, Russia," *GeoJournal* 78(5): 817–32.

Graybill, Jessica and Megan Dixon. 2012. "Cities of Russia," in S. Brunn, M. Hays-Mitchell, and D. Ziegler (eds), *Cities of the World: World Regional Urban Development*. Fifth edition. Lanham, MD: Rowman and Littlefield.

Henry, Laura A. 2010. *Red to Green: Environmental Activism in Post-Soviet Russia*. Ithaca, NY: Cornell University Press.

Henry, Laura A. and Lisa McIntosh Sundstrom. 2007. "Russia and the Kyoto Protocol: Seeking an Alignment of Interests and Image," *Global Environmental Politics* 7(4): 47–69.

Hoornweg, Daniel A., Mila Freire, Marcus J. Lee, Perinaz Bhada-Tata, and Belinda Yuen (eds). 2011. *Cities and Climate Change: Responding to an Urgent Agenda*. Washington, DC: World Bank.

IEA (International Energy Agency). 2010. *World Energy Outlook 2010*. Retrieved 10 July 2014 from http://www.worldenergyoutlook.org.

IPCC. 2014. *Climate Change 2014: Impacts, Adaptation and Vulnerability. Fifth Assessment Report*. Geneva: Intergovernmental Panel on Climate Change.

Josephson, Paul, Nicolai Dronin, Aleh Cherp, Ruben Mnatsakanian, Dmitry Efremenko, and Vladislav Larin. 2013. *An Environmental History of Russia*. Cambridge: Cambridge University Press.

Korppoo, Anna. 2008. "Russia and the Post-2012 Climate Regime: Foreign Rather than Environmental Policy." Briefing Paper 23, 24 November. Helsinki: Finnish Institute of International Affairs.

Krasheninnikov, Stepan P. 1972. *Exploration of Kamchatka*. Trans. E.A.P. Crownhart-Vaughan. Unabridged edition. Portland: Oregon Historical Society.

Lioubimtseva, Elena. 2010. "Russia's Role in the Post-2012 Climate Change Policy: Key Contradictions and Uncertainties," *Forum on Public Policy*, no. 3: 1–11.

Loboda, T. 2014. "Adaptation Strategies to Climate Change in the Arctic: A Global Patchwork of Reactive Community-Scale Initiatives," *Environmental Research Letters* 9: 111006.

McLaughlin, Paul and Thomas Dietz. 2007. "Structure, Agency and Environment: Toward an Integrated Perspective on Vulnerability," *Global Environmental Change* 18: 99–111.

Morton, Henry and Robert Stuart (eds). 1984. *The Contemporary Soviet City*. Armonk, NY: M.E. Sharpe.

Moser, Suzanne. 2010. Communicating Climate Change: History, Challenges, Process and Future Directions *WIREs Climate Change* 1: 31–53.

Oldfield, Jonathan. 2013. "Climate Modification and Climate Change Debates Among Soviet Physical Geographers, 1940s–1960s," *WIREs Climate Change* 4: 513–24. doi: 10.1002/wcc.242.

Pelyasov, Alexander, 2009. "The Arctic in the New Creative Age: The Arctic Dimension of the Knowledge Economy," in UNESCO (ed.), *Climate Change and Arctic Sustainable Development: Scientific, Social, Cultural and Educational Challenges*. Paris: UNESCO.

Petrov, Andrey. 2011. "Beyond Spillovers: Interrogating Innovation and Creativity in the Peripheries," in H. Bathelt, M. Feldman, and D.F. Kogler (eds), *Beyond Territory: Dynamic: Geographies of Innovation and Knowledge Creation*. London: Routledge, pp. 168–90.

Pryde, Philip. 1991. *Environmental Management in the Soviet Union*. Cambridge: Cambridge University Press.
Pugliese, Anita, Julie Ray. 2009. A Heated Debate: Global Attitudes toward Climate Change. *Harvard International Review* 31(3): 64–68.
Putin, Vladimir V. 2010. Address to the International Arctic Forum. Retrieved 7 July 2014 from http://int.rgo.ru/news/vladimir-putin percentE2 percent80 percent99s-address-to-the-international-arctic-forum/.
Rosenzweig, Cynthia, William D. Solecki, Stephen A. Hammer, and Shagun Mehrotra. 2011. *Climate Change and Cities: First Assessment Report of the Urban Climate Change Research Network*. Cambridge: Cambridge University Press.
Rowe, Elana W. 2009. "Who Is to Blame? Agency, Causality, Responsibility and the Role of Experts in Russian Framings of Global Climate Change," *Europe-Asia Studies* 61(4): 593–619.
Russian Geographical Society. 2011. "Vladimir Putin Distributes Grants at the Meeting of the Russian Geographical Society's Board of Trustees." Retrieved 3 June 2012 from http://int.rgo.ru/news/vladimir-putin-distributes-grants-at-the-meeting-of-the-russian-geographical-society's-board-of-trustees/.
Sabonis-Helf, Theresa. 2002. Climate Change Policy in Russia, Ukraine and Kazakhstan. Washington, D.C.: *The National Council for Eurasian and East European Research*.
Sclar, Elliott D., Nicole Volavka-Close, and Peter Brown (eds). 2013. *The Urban Transformation: Health, Shelter and Climate Change*. New York: Routledge.
Sharakhmatova, Viktoria. 2011. *Observations of Climate Change by Kamchatka Indigenous Peoples*. Report for Lach Ethnoecological Information Center. Petropavlovsk-Kamchatsky: Kamchat Press.
Sharkun, V.V. 2008. *Kamchatka, People, Climate*. Petropavlovsk-Kamchatksy: New Books.
Shiyatov, Stefan G. 1988. "The Development and State of Dendrochronology in the USSR," *Tree Ring Bulletin* 48: 31–40.
The Russia Monitor. 2012. "New Law on 'Foreign Agent' NGOs." Retrieved 13 July 2013 from http://therussiamonitor.com/2012/08/02/legal-update-new-law-on-foreign-agent-ngos/.
Turkowski, A. 2012. *Russia's International Climate Policy. Polish Institute of International Affairs Policy Paper*. Retrieved 8 December 2015 from https://www.pism.pl/files/?id_plik=10025.
Tynkkunen, N. 2010. "A Great Ecological Power in Global Climate Policy? Framing Climate Change as a Policy Problem in Russia Public Discussion," *Environmental Politics* 19(2): 179–95.
UN Development Programme in Russia (UNDP). 2009. "Integrated Climate Change Strategies for Sustainable Development of Russia's Arctic Regions (Case Study for Murmansk Oblast)." Moscow: UN Development Programme in Russia, Russian Regional Environmental Centre.
World Resources Institute. 2015. "Russia's New Climate Plan May Actually Increase Emissions." Retrieved 20 December 2015 from www.wri.org.

CHAPTER ELEVEN

Conclusion
Drivers of Change
Robert W. Orttung

The contributors to this volume have laid out some of the key challenges facing Russia's Arctic cities in terms of promoting sustainability and started to sketch their responses to these challenges. The challenges involve developing the resources located in Russia's Far North with the least impact on the natural environment and strengthening the social fabric of the cities there. In dealing with these issues, Arctic city managers are addressing decisions made by federal authorities in Moscow, the country's prioritization of resource development and the resulting manpower needs to achieve this objective, and the influence of climate change. Among the responses this volume describes are efforts to diversify the monoprofile economies of important cities through the development of small business, discussions surrounding immigrants to the Arctic cities, and measures designed to deal with thawing permafrost through improved building techniques.

The chapters in this book have focused on three drivers of change: policy making, resource development, and climate change. Now it is possible to pull together some of these findings into a broader synthesis.

Policy Making

Several of the chapters observed that policy makers in Moscow shaped Russia's Arctic urban policy with little regard for issues of sustainability in both the Soviet and post-Soviet periods. While there is no denying that President Vladimir Putin has played a significant role in the process for the last fifteen years, a focus exclusively on his personality and preferences is far from sufficient to understand the nature of Russia's urban policy in the Arctic. As Elana Wilson Rowe demonstrated in Chapter 2, a variety of federal agencies and experts participate in the debate about Arctic policy. But, even so, input from regional governments and indigenous organizations did not have much influence on the federal-level decisions she examined. Moreover, the Arctic debates she analyzed included little discussion of climate change. Likewise, Jessica Graybill's analysis in Chapter 10 of perceptions of climate change among various Arctic populations showed that they did not see this issue as personally threatening nor did they accept Western notions of climate change. As a result, neither federal nor regional policy seems to be addressing the issue with much vigor.

According to Scott Stephenson's Chapter 8, national development priorities will drive growth in the Arctic, as they have in the past. While climate change will serve as an important backdrop to the realization of these plans and have a big impact on what happens in the Arctic, the preferences of Moscow policy makers will determine what kind of actions Russia takes in the North. Whether such policies are viable in the future, as the impact of climate change becomes more apparent, remains to be seen.

Wilson Rowe and Marlene Laruelle (Chapter 5) noted that the predominant topic of debate among Moscow policy makers was whether Russia should pursue open policies aimed at promoting its economy or whether it should emphasize greater closure to ensure its security objectives. While the discussion has not reached any definitive conclusions, Stephenson argued that Russian behavior regarding potential shipping through the Northern Sea Route indicated a de facto prioritization of increased economic activity.

Such a focus on openness and the integration of Russia into global commodity flows was a break from past Soviet practices. However, the future of such open policies remained unclear after Russia invaded Ukraine and annexed Crimea, leading the West to impose economic sanctions targeting Russia's economic elite and technology exports that would support the development of Russia's Arctic oil reserves. Putin's initial response was to announce policies designed to reduce Russia's dependence on the outside world. While it is still too early to tell what the full impact of the Ukraine conflict will be overall, the conflict will certainly affect the way that Russia seeks to define its place in the world with clear implications for what happens in the Arctic.

Resource Development and Labor Flows

The Russian economy's continued reliance on natural resources, and particularly oil and gas, is a given for the near- to medium-term future. The importance of this industry in the Arctic is already apparent in the flows of workers to the region. Timothy Heleniak's overview in Chapter 4 shows that the main cities growing in the Arctic are connected to resource development. Laruelle examined the composition of the new migrants moving to the Arctic and explored how they are being integrated into existing Arctic communities. Finally, Gertrude Saxinger, Elena Nuykina, and Elisabeth Öfner explained in Chapter 6 how shift workers are providing links between the resource industry of the north and some of Russia's southern regions.

These chapters raise serious issues that require further study. Looking at the Arctic cities, it is necessary to understand transformations of the urban landscape. As Laruelle points out in the conclusion to her chapter, such transformations in resource-driven cities include new social stratification in terms of capital, values, and identity and new cultural interactions between "urban dwellers," "indigenous," and "newcomers." It will also be necessary to understand how members of various ethnic groups are integrating into existing cities and what impact the establishment of compact ethnic quarters will have on these urban areas. Another area of interest is how the corporations active in the Arctic interact with the city authorities and residents.

In addition to the cities, it is necessary to focus on the workers who are moving to these cities. To understand the future development of the cities, workforce and migrants' own professional strategies will play a key role, as Laruelle notes. Similarly, as Chapter 6 points out, shift workers often face deteriorating working conditions as their employers seek to maximize profits in an overall system defined by corruption and dependent courts. The authors point out that existing working conditions are frequently in violation of Russia's labor code. Rather than resolving these issues, the state lacks a clear policy to enforce proper conditions on energy sector companies that now employ hundreds of thousands of people.

All of these trends raise questions about the ability of Russia's Arctic cities to address sustainability issues. Moving forward there is even uncertainty about the role that Arctic energy resources will play in Russia's overall economic development. Energy industry experts like Valeriy Kryukov and Arild Moe pointed out in 2013 that "it is still unjustified to state without reservation that the Russian Arctic continental shelf is set to become a major arena for the international oil industry in the course of the next ten years" (Godzimirski 2013). The global energy markets remain extremely volatile, and this uncertainty will likely affect the development of Russia's energy industry and the Arctic cities that supply its labor.

Climate Change

Finally, Russian scientists have long been aware of climate change and have contributed extensively to international understandings of this topic. Nevertheless, this deep expertise does not always translate into informed decision making about how to address the issue. While climate change has had a big impact on transportation links and infrastructure, neither federal nor regional policy makers have provided active responses.

The chapters in this volume draw attention to urban development associated with the expansion of shipping along the Northern Sea Route and the impacts of warmer climates for reducing the season when ice roads can be employed (see especially Chapters 7 and 8). The contributors also provided a quantitative measure of the impact of permafrost thawing on urban and energy sector infrastructure (Chapter 9). These challenges will certainly make up some of the key issues facing Russia's Arctic cities moving forward. As Graybill's chapter points out, understanding the vulnerabilities of the different cities in a detailed way will be crucial to identifying measures that can address such issues.

Future Research Directions

As noted in the Preface, the contributors to this volume see the chapters gathered here as a first step in addressing Arctic urban sustainability in Russia rather than a definitive analysis of the problem. So far, we have examined several of the key factors influencing Russia's Arctic cities and their efforts to achieve sustainability. There is still a considerable amount of work to be done. Having looked at the broader forces affecting the cities, it is now possible to increase the level of granularity to the city level.

First, there are only a few works on the "black box" of Russian policy making for the Arctic (Wilson Rowe 2011). This topic deserves a lot more attention so that we can development a better understanding of how the policy-making process works and how actors who support sustainability goals can have a stronger influence on it. Currently, the actors with the greatest interest in promoting sustainability—local governments and nongovernmental organizations—have the least amount of impact on the actual policy process, while the actors making decisions—currently the Kremlin and large state-owned corporations—have priorities other than sustainability (Orttung 2015). Improving the state of knowledge about Russian decision-making processes for Arctic urban development at both the federal and local levels will help contribute to the broader political science literature on decision-making in Russia.

A second approach that would build effectively on the current volume is to develop case studies of several representative cities in the Russian Arctic, explaining how they can increase their level of sustainability through improved political, economic, design, planning, and other measures. Such case studies could focus on the issues raised in this volume, but delve into much greater detail. The study of Gubkinsky and Muravlenko presented in Chapter 3 by Nadezhda Zamyatina and Alexander Pelyasov showed how one city was able to outperform the other across a wide range of social indicators, spanning from small business to museum attendance. Although part of Gubkinsky's success derived from its location farther away from the controlling influence of Noyabrsk, its leaders' ability to diversify a monoprofile city provides lessons for policy makers. More case studies of successful and struggling cities could provide new insights into figuring out what works.

Case studies of Norilsk and some of the energy-focused cities of the Yamal-Nenets Autonomous Okrug could examine how workers are integrated into the urban fabric of the city and how such efforts promote or hinder sustainability. Other studies could focus in on the treatment of workers and highlight situations that promote both worker welfare and corporate advances. Improving the conditions of workers would contribute to high levels of sustainability.

Studies focusing on how cities adapt to climate change likewise could help identify paths forward. Particularly important are case studies of cities addressing transportation and building infrastructure issues.

Moving beyond the topics addressed in this volume, city-level studies could also address issues of housing, education, and health care provision in Russia's Arctic. Such analyses would provide valuable new insights to scholars working on these questions and practical guidance to policy makers.

Finally, in a third area for further research, case studies that place Russian Arctic cities in comparative international perspective could shed new light on what is happening in Russia and provide potential new ideas on how to proceed based on best practices from around the world. Such cases could be drawn from other Arctic countries, such as Norway, Canada, and the United States, or even non-Arctic cities that face similar problems of harsh climates, fragile environments, boom-bust economic cycles, and extensive migration, such as Phoenix, Arizona.

This volume has sought to define the key forces shaping Russia's Arctic cities in an effort to understand their current levels of sustainability and potential for further improvement. In doing so, it sets the stage for additional research addressing the way that Russia's Arctic cities deal with the challenges they are facing.

Robert W. Orttung is Research Director for the George Washington University Sustainability Collaborative and Associate Research Professor of International Affairs.

References

Godzimirski, Jakub M. (ed.). 2013. *Russian Energy in a Changing World: What Is the Outlook for the Hydrocarbon Superpower?* Farnham, UK: Ashgate.

Orttung, Robert W. 2015. "Promoting Sustainability in Russia's Arctic: Integrating Local, Regional, Federal, and Corporate Interests," in Susanne Oxenstierna (ed.), *The Economics of the Russian Politicised Economy: Institutional Challenges Ahead.* London: Routledge.

Wilson Rowe, Elana. 2011. "Encountering Climate Change," in Julie Wilhelmsen and Elana Wilson Rowe (eds), *Russia's Encounter with Globalization: Actors, Processes and Critical Moment.* Houndmills, UK: Palgrave Macmillan.

Index

Abramovich, Roman 72
accessibility, and economic development 176; and effect of air temperature 203; rail 33, 48, 90, 91, 118; road 90, 176; sea 19, 20, 90, 176, 177; transit 12, 18, 19; water 158
adaptive capacity 222, 234, 236
Afghanistan 94
agricultural sector 74
Alaska 33; economy 52; and ice roads 189; and oil and gas 178, 184–6; rail link to Russia 33; urbanization 1
albedo 214
Alrosa 91
Amur 188
Anadyr 78, 209, 214; and food security 234
Apatity 79, 84
Arctic amplification 18, 150
Arctic Commission 28
Arctic Council 36
Arctic Human Development Report 69
Arctic policy 26, 27, 28, 30, 38
Arctic security 28, 30
Arctic states 179
Arkhangelsk 69, 70, 90, 91; administrative center 79, 233; and ice jam floods, 163; innovation, 234; port 5; population changes 72, 78, 80, 84
Armenia 96, 98

Avacha Bay 92
Azerbaijan, immigration 94, 96, 98, 101, 102; skilled labor 126
Azeris 96, 102

Barents Euro-Arctic Transport Area (BEATA) 90
Barents Sea 25, 90
Barents Sea boundary agreement (2010), 193
Bashkortostan 100, 116; and LDC 119, 120, 126–8; and education 126–7, 129; dependency on oil and gas sector 128
Belarus 96
Belaya Gora 214
Belkomur railway project 91
Berezovsky, Boris 49
Bering Sea 33
Bering Strait 179
Beringovskiy, infrastructure instability 213
Bilibino 211
Bovanenko 91, 123
Bovanenko-Ukhta Pipeline 116, 120
Bovanenkovo Gas Field 17
Budyko, Mikhail, 143
Buryat Republic 34

Canada 32; accessibility 188–90; emigration 94; international cooperation in Arctic 32, 179, 245;

oil and gas 178; permafrost 203; sustainable development in the Arctic 31; urbanization 1
carbon credits, Russia 225
Carbon Disclosure Project for Russia (2009) 227
Central Asia 94, 96, 99, 102, 106
central authority 14
central planning, Soviet 70, 80
Chechnya 99
Cherskiy 211, 214
China 48, 91, 92; and immigration 94
Chukotka AO 68–69, 78, 149, 163; climate change 149, 159, 162-3, 209–211; economy 91–92; permafrost 213; population change 72–73, 85, 92, 209–10, 234; resettlement 93
climate change 2, 14, 36, 38, 141–172; accessibility 175–6, 187, 243; adaptation 35; Arctic development 195; corporate involvement 227; energy demand 158, 163; Far North cities 144; fog and reduced visibility 184; health 158; indigenous knowledge of 230, 236; permafrost 20, 184, 201–4, 207, 212–14; population 209, 210–12; positive feedback 180; public perception 222, 228, 230–31; Russian policy 20, 235; Russian science 243; sea ice 19, 143, 184; socioeconomic consequences 203; urban vulnerability 216, 221–236
climate groups, international, Russia's involvement 228
climate leadership, Russia 222
climate policy talks, international, Russia's role 224
climate policy, Russian North 226, 230
climate projections 150, 151
climate science, Soviet 223
coal 7–9, 11; and permafrost 215; and subsidies 11
coal mining 12, 133, 234; and economy 85, 116. *See also* Vorkuta, Pechora coal basin

collective memory 53
Commonwealth of Independent States (CIS) 89, 94, 96, 100; migrants 99, 106, 108
construction industry 102, 125
Crimea 26, 38, 241
crude oil 29, 126

Dagestan 101
dairy industry 43, 44
deindustrialization 8, 11, 16
delimitation line 25
demographic decline 11, 14, 19
demographic growth 34
depopulation 11, 13
Deputatskiy 214
diamond industry 84, 91, 168
diaspora 48
Dikson 93, 214
diversification, economic 60, 79, 121, 122
dual state (Russia) 27
Dudinka 78, 91, 192, 210

economic development 38, 113; institutional factors 45, 46; local level 42
economic exploitation 34
economic geographical situation 42, 45, 47, 48, 49
economic growth 28
economic policies, openness 241
economic transition, Arctic, post-Soviet 76
economy, Arctic 34; natural resource-based 8, 9
economy, Russian, and dependency on Arctic 37, 88; transition to market-based 70
education 33, 56, 78, 105, 244; indigenous peoples 234; petroleum-related universities 125, 127. *See also* Bashkortostan
Egvekinot 211
emigration, Russian 94
energy center 13
energy development 2, 14, 15, 16
energy exploitation 4, 12
energy exploration 13, 14, 16

energy industry, and economy 13, 14, 16, 91, 242–3; and urban growth 17; and politics 18, 26
Engels Dictum 4, 7
environment 35, 74, 88
ethnic homelands 76
European Arctic 90
European North, urbanization 75
Evenki AO 68, 69, 73
extractive industries 102

Far Eastern District 91
Far North 68–69; accessibility 150; housing 55; and climate change 18–20, 144, 212; economic development in Russia 1, 8, 10; and emigration 92; and energy exploitation 15; and forced labor 4–5; population 11, 92, 112, 123; pre-Soviet development 3; Soviet industrialization 4, 7, 11, 17; urbanization 1, 10
Federal Migration Service 95, 97, 101
Federal Security Service, Russia 30, 37
Federal Service for Hydrometeorology and Climate Monitoring 235
Finland 90, 91
fishing industry 90, 92
flood, Lena River 158
forced labor 29
Franz Josef Island 25
Fundamentals of State Policy of the Russian Federation in the Arctic in the Period up to 2020 and Beyond (2008) 32

Gazprom 13, 15, 17, 100, 125; climate change 227; labor 120; monopoly 192; pipelines 119
Gazprom Dobycha Noyabrsk 54
Gazprom Neft 188
Georgia 94
Glavtyumenyeftegaz 47
gold 91
Gorodilov, V.A. 46
Great Circle Route 179

Greenland 32; international cooperation in the Arctic 32, 179; oil and gas 178; urbanization 1
Gubkinsky 43–60
Gubkinsky City Administration 52
Gubkinsky Council of Entrepreneurs 43
Gubkinsky Museum of the Development of the North 53
Gulag 3, 4, 5, 7, 10; and rail development 91; and Arctic development 206

Hickel, Wally, 52
housing, rising cost 120
housing surplus 74
hydrocarbons 43, 91, 114–5, 191; and professional qualifications 104; as energy source 178; and economy 186
hydropower 8, 158

ice jams, flood risk 157, 163, 169
ice roads 91, 176, 187, 188; reduced usability 158, 189; and role in economic development 188
icebreakers, private corporations 194
identity, Arctic 93
Igarka 5, 20, 93
Illarionov, Andrei, 142
Illullissat Declaration 32
immigration quota 95, 96
indigenous organizations, impact on federal policy 241
indigenous people 2, 35, 94, 107; birth rate 93, 100; and climate change 222, 236; cultural preservation 33; and the environment 31; living conditions 94; migration 13; and protection of livelihood 34, 38; and urban settlements 210, 233, 234; and understanding of climate change 229; in urban areas 209
industrial development, post-WWII 7, 8, 206
industrial production 89

industrialization 5, 17, 74, 78; Gulag era 3–4
infrastructure 68, 88, 90; investment in 89; labor demands 107; transport 17
Institute of Economic Analysis (IEA) 142
integration of migrants 106
Intended Nationally Determined Contribution (INDC) 225
Interagency Commission 227
Intergovernmental Panel on Climate Change (IPCC) 143, 144, 150
Irkutsk 34, 91, 169
Israel 48, 94
Italy 48
Izrael, Yuri, 224

Japan 92

Kamchatka 73, 84
Kamchatka Kray 69, 78, 85
Kamchatka Oblast 68
Kamchatka Peninsula 92; and climate change 229, 230
Karelia 69, 70, 72; population changes 78, 84, 90
Kayerkan 91
Kazakhstan 94, 96, 104
Khanty-Mansiy AO 8, 69, 70, 84, 126; and oil and gas 74, 79, 80 117; and migration 72, 91, 101, 102; and population 73, 74, 76
Khanty-Mansiysk 100
Khanymei 48
Khrushchev, Nikita 206
Kirill 43
Kogalym 79, 126
Kola peninsula 192
Kolyma 5
Kolyma basin 91
Kolyma Federal Highway 188
Komi Republic 69, 70, 91; population 78, 79, 80, 84; and coal, oil, and gas extraction 85, 113
Koralym 80
Korea 94
Koriak AO 68, 69

Kotlas 84
Krasnodar, migration 99
Krasnoyarsk 2, 79, 84, 91; and migration 100; and youth 93
Krasnoyarsk Kray 7, 34, 69, 78; population 7, 210; labor force 9; migration 123
Kremlin 15, 16, 25
Kyoto Protocol, 224
Kyrgyzstan, migration 94, 96, 101, 102

labor, black market 101; forced 1, 4, 5, 7, 9; shedding 11; shortage 9; wild commuting 125, 132
Labytnangi 17, 59, 188
Latin America 48
Lavrov, Sergei, 36
Lena 19
Lena Federal Highway 188
Lena River 169, 188
Leningrad 3
Leningradsky, gold mining 211
Lensk 158, 214; and flood 168–171
liquid natural gas (LNG) 179, 192
local authority, political 60
local identity 59
long distance commute (LDC) system 113, 115, 116, 132; and changing community life 115; and gender 123; impact on communities 115, 116, 120; and increased social-spacial proximity 115; and rising crime 124; and urbanization 127
long distance commute (LDC) workers 117, 119, 128, 129, 130, 133; economic threat 121; ethnic groups 129; and housing 116, 120, 124, 127; as impetus for development 121; and incidence of STDs 122; and increased income 127; increased socioeconomic standing 126; and prostitution 122; and real estate sector 122

Magadan 5, 29; population changes 72, 75, 78, 84, 92–3; transportation 188; urbanization 201

manpower problem 8, 9,
Maritime Doctrine, Russian
  Federation 32
Matevosov, Aleksei 49, 52
Medvedev, Dmitry 25, 30, 37
Medvezhye 8
metallurgy 104
migrant, illegal 100, 101, 105
migration 70–73, 75; and collapse
  of the Soviet Union 92; illegal
  95, 97, 98, 101; internal 93, 99;
  labor-related 11, 13, 100, 102, 141;
  long-term 107; resource-driven 20;
  Russian policy 95, 107; and Siberia
  4, 10; Slavic 2; urban 84, 85, 89,
  92; youth 100
Mikhailov, M.K. 46
military installation 2, 29
mineral resources 8, 89, 93, 187;
  and economy 91, 114, 217; and
  industrialization 92, 112, 206;
  and migration 94; *See also* natural
  resources
mining industry 90, 121
Ministry for Regional Development
  28, 30, 37
Ministry of Foreign Affairs 36, 37
Ministry of Natural Resources 30, 226
Ministry of Oil Industry 47
Mirnvi, diamonds 168, 169
Moldova, immigration 94, 98
Monchegorsk 79, 84
monocities 11, 15, 29, 42–61, 124
monogorody *see* monocities
mortality 70
Moscow 3, 15, 16, 17; and central
  authority 29; and metropolitan
  economic center 76
municipal authority 49
Muravlenko 42–60, 213
Muravlenko Ecological-Regional
  Studies Museum 53
Muravlenkoyskneft 47
Murmansk 29, 68, 77, 90; population
  70, 78–80, 84, 93; as transit hub
  12, 74, 192
museum affairs 53, 54, 59
Mys Schmidta 211

Nadym 13
Nar'yan Mar 78, 100, 191
National Center for Environmental
  Prediction 208
natural gas 8, 9, 29, 123, 126, 234;
  as fuel for ships 179; and Soviet
  economy 9; and Yamal Peninsula
  120, 130
natural resources 4, 29, 36, 74, 88;
  and economic importance 30,
  38, 67, 112, 242; and sustainable
  development 31. *See also* mineral
  resources
Neelov, Yuri, 16
Nefteyugansk 17, 76, 79, 80
Nenets AO 68, 78, 91, 208; and
  climate change 210, 212; and
  indigenous people 210; and
  population growth 73
neoliberalism, restructuring of the
  north 113–4
Neryungri 84, 211, 214
nickel 30
Nizhnevartovsk 76, 79, 84, 126
Nogliki 234
Norilsk 5, 13, 15, 69, 78–80; and
  climate change 211, 214–16; and
  migration 100, 103; resources 7,
  91; and permafrost 20, 206, 210
Norilsk Nickel 13, 91, 192, 194; and
  migration 104; and relationship
  with the state 15
North Caucasus 99
North Pacific trade 92
North Sea Fleet 74
Northern (Arctic) Federal University
  84
Northern Benefits 9, 10, 89, 90, 92
Northern Fleet 12, 90
Northern Policy 29
Northern Railway Company 121
northern resettlement 131
Northern Restructuring project 93
Northern Sea Route (NSR) 12, 90,
  91, 158, 184; and accessibility
  143, 179–180; and climate change
  18, 19, 150, 176; and economic

development 29, 32–33, 35, 177, 187, 191–3
Northern Shipment 8, 9, 11, 19
Northwest Passage 180
Northwestern Federal District 90, 91
Norway 32, 91; and research vessels 36
Norwegian Lapland 90
Novatek 15, 17
Novaya Zemlya 90
Novokuznetsk 100
Novoportovskoye field 188
Novosibirsk 2, 100
Novyy Urengoy 9, 13, 17, 59, 201; and climate change 211, 214; and development 126; and indigenous people 210; long distance commute (LDC) 116–17, 12–126; and oil and gas 76, 79; population 84, 130
Noyabrsk 45–50, 59, 76; and climate change 210–213; and oil and gas 79–80
Noyabrskneftegaz 46, 47
Noyabrskneftestroy 46
nuclear power plants 90, 100
Nyagan 79, 84

Ob 19
Ob-Bovanenko rail line 17, 91
occupational health and safety 100, 104, 105
Odessa 4
Office of Visas and Registration 95
oil and gas 43, 178, 186, 191; Russia's reliance on 242
oil and gas development 183; Arctic 32, 176, 183, 184, 186, 191; physical and economic challenges, 176
oil and gas exploitation 100
oil and gas extraction 53, 89
oil and gas fields 8, 117, 119, 178; economic opportunity for youth 93; Usinsk and Vyktyl, 119
oil and gas industry 9, 16, 84, 193, 213; changing perceptions 126; and employment 116, 119, 127–128; and migration 100; and population 56, 92; and Soviet economy 17; and sustainable development 31
oil and gas regions 70, 74, 76, 80, 178; and population 79, 100
oil deposits, Siberia 103
oil spill, environmental impact 184, 185
Omsk 34, 100
Oslo 25
overpopulation 34

Passive Principle 207
Pechora coal basin 5
Pechora Sea, proximity to oil fields 191
permafrost 18, 20, 100, 205, 208; engineering 212; thaw 91, 141; and climate change 201, 204; and construction methods and norms 206, 207, 208; discontinuous 74; and effect of air temperature 160; and effect of snow 215; and effect of vegetation 216; and effect on infrastructure 157, 184, 202, 204
permanent settlement 34
Petropavlovsk-Kamchatskiy 78, 79, 80, 84–5, 92
Petrozavodsk 78, 80, 84
Pevek 211, 214, 234
Plesetsk Cosmodrome 90
population, boom 5; change 67; crisis 7; decline 12–13, 74, 84, 89, 92–93; growth 56, 70, 80; natural decrease 70, 72, 75, 84; natural increase 72, 75, 85; and post-Soviet collapse 68, 70, 192
post-Soviet transition 8, 11
poverty gaps 93
Presidential Administration 37
Primorye 91
Prokhorov, Mikhail 15
Provideniya 213
Purneftegaz 47, 49, 50
Purpe 48
Putin, Vladimir, and Arctic cooperation 36; and Arctic security 30; and centralization 29; and

energy sector 15; and geopolitical interests 25; and policies 15, 16; economic importance of Northern Sea Route 183; and rebuilding of Lensk 168; and rise to power 14; and sensationalism 142; and strategic importance of Arctic 112

regional development 42
regional government 29, 38, 241
restructuring policies, post-Soviet 130
Rogozin, Dmitry 28
Rosatom 100
Rosatomflot 194
Rosneft 15, 17, 47, 49
Russia, and border control 25; and climate change 146, 229; and climate policy-making 243; and economic development 1, 2, 29, 32, 67; and economic interests in Arctic 25–26; and energy strategy 32
Russia, and environmental issues 30; and military infrastructure 25; national security, 26, 32, 33, 36; reliance on oil and gas 242; sustainable development 31, 33, 67, 244
Russian Civil War 3
Russian Empire 2
Russian Far East 85
Russian Hydrometeorological Service 227
Russian parliament and climate change 229
Russian Revolution of 1917, 3
Russian Security Council 107

Sabetta 17
Sakha Republic 68, 209, and climate change 169, 210, 212–214, 229; and infrastructure instability 213; and population changes 73, 79, 80, 84, 209. *See also* Yakutia
Sakhalin 69, 73, 78, 79, 84
Salekhard 59, 201, 214; and accessibility 188; and migration 100; oil and gas 13, 17, 76
Salekhard-Igarka railway 91

Samotlor 8
sanctions, against Russia 26
sea ice 184; reduction 150, 176, 180
Security Council, Russian 36
security, joint, Russia and Nordic countries 90
service sector 102
Severniy 214
Severodvinsk 78, 80, 84
Severomorsk 79, 84
shift work 17, 34, 38, 104
Siberia, bridge between Europe and Asia 38; and development 1–5; early settlement of 88; geography 33
Siberian Federal District 91, 93
Sibneft 49
small business 43, 44, 59, 60
snow, importance of 215; effect on permafrost 159, 215; effect on water levels 159; thermal insulation 215
snowfall 160
social stratification 89
social ties, long-term 129
soil salinization 215
South Caucasus 94, 99
South Korea 92
Soviet Arctic industrial cities 30
Soviet Union, and agriculture 80; and central planning 67, 223; and environmental interests 31
St. Petersburg 80, 90, 99
State Hydrological Institute, Russia 143
state involvement in regional policy 114
state subsidies 29
Stoltenberg, Jens 25
Strategy for Socio-economic Development of Siberia towards 2020 (2010) 32
Strategy for the Development of the Arctic Zone of the Russian Federation (2013) 32
Stroygazkonsulting 120, 125
Surgut 9, 17; and energy 13, 15; and LDC 126; and oil and gas 79; and population changes 80, 100–101

sustainability 74, 242; social 73, 88–108; urban 14, 16, 20, 74; urban Arctic 86, 88; and workers 244
sustainable development 31, 35; and oil and gas development 31
Syktyvkar 78, 80

Tajikistan, migration 94, 96, 101, 102
Tako-Sale 213
Talnakh 91, 210
Tashkent Agreement (1992) 94
Tatarstan 100
Taymyr AO 69, 78, 90–92, 214
Taymyrsky Dolgano-Nenetsky District 208, 210. *See also* Taymyr AO
Tazovskoye field 178
territorial identity 52, 53, 57
Tiksi 93, 191
Tomsk 34
tourism 91
trade unions 50
Trans-Siberian Railroad 3, 5, 7, 91, 206
Transpolar mainline 91
transportation 34, 48, 84, 92; public 86
Tsarist period 2, 90
Turkey 48, 94
Turkmenistan 94
Turuchansk 123
Tyumen 91, 100, 102

Udmurtia 100
Ukhta 79, 84, 85
Ukraine, immigration 94, 96, 102
United Nations Climate Change Conference in Paris (COP21) (2015) 225
United Nations Framework Convention on Climate Change 225
United States 32, 75, 94; military build-up 36
University of the Arctic 36
Ural Federal District 9
Ural mountains 90
urban development, Arctic 26; Russian Arctic, natural resource-driven 155

urban sustainability 141, 159
urbanization 1, 3, 74, 84; Arctic 88, 201; and GDP drain 19; global trend 88; state-driven 3
Ursinsk, oil and gas fields 119
Ust-Kuyga 214
Uzbekistan, immigration 94, 96, 98

Vankor 123
Vietnam, immigration 94
Vladivostok 92
Vorkuta 79; and climate change 20, 201, 214–15; 79; and coal 7, 117–118, 130, 131, 133; and Gulag 5, 118; and LDC 119–124; and population changes 84–85, 100, 116, 118, 234
vulnerability, urban areas in Russian Arctic 231
Vyktyl, oil and gas fields 119

West Siberia 70, 75
White Sea-Baltic Canal 4
working conditions 100, 125, 244
World Bank, and reports on cities' role in economy 85–86, 93, 118
World Climate Change Conference in Moscow (2003) 224
World Council of Mayors and Cities Climate Leadership Group (C40) 228
World Mayors Council on Climate Change 228
World Resources Institute 225
World War II 5, 7, 8

xenophobia 107

Yakutia 68, 70, 78, 91, 171; and climate change 160, 162, 169, 170; and economic decline 92; and permafrost warming 205; and population changes 84, 93. *See also* Sakha
Yakutsk 70, 100, 188, 201; accessibility 12; and climate change 20, 205, 211; and diversification 13–14; and

migration 92; and population 78–80, 84, 88, 92
Yamal-Nenets AO 15, 43, 68, 91, 188; and climate change 211; and gas extraction 112–3; and infrastructure instability 213; and long distance commute workers 123; and migrants 102; and oil and gas 8, 13, 79, 112–113, 117; and permafrost 208, 213; and population change 70, 72–76, 80, 84, 92–93, 100; and sustainability 16; and urban population 209–210; and urbanization 75; and workforce development 115, 244
Yamal Peninsula 119, 120, 123, 188, 211; and climate change 229; and oil and gas sector 119
Yamburg 8, 123
Yekaterinburg 36; and migration 99
Yenisei 19
youth 56, 70, 84, 100; and economic prospects 92; and migration 93
Yuzhno-Sakhalinsk 78, 79, 80, 84

Zapolyarnoe 123

www.ingramcontent.com/pod-product-compliance
Lightning Source LLC
Chambersburg PA
CBHW070915030426
42336CB00014BA/2422